15 FEB 1974

MARKET WEIGHTON SECONDARY SCHOOL
EAST RIDING EDUCATION COMMITTEE

Market Weighton
School Library
Withdrawn

978 0853401681

D1785287

Books are to be returned on or before the last date below

27 MAR 1974

16 DEC 1974

21 MAY 1975

26 FEB 1976

-5 MAR 1976
- 6 MAR 1978

-2 DEC 1980

Market Weighton
School Library
Withdrawn

LIBREX —

573

5917

TRACKING FOSSIL MAN

Excavating a cave in France. *Photo courtesy Carlyle S. Smith.*

Tracking Fossil Man

AN ADVENTURE IN EVOLUTION

Sharon and Thomas McKern

WAYLAND PUBLISHERS LONDON

SBN 85340 168 3

This edition first published in 1972
by Wayland (Publishers) Limited
101 Grays Inn Road, London WC1
Copyright © 1970 by Praeger Publishers, Inc.
Printed photolitho in Great Britain by
J. W. Arrowsmith Ltd., Bristol

For Rosa McPhail
with gratitude and deep affection

Contents

vii

List of Illustrations

Preface

Man is a very special animal. He is endowed with a large and highly developed brain, a pair of dexterous hands, the power of speech, and a unique capacity for culture—qualities that permit him to survive and flourish in masterful control over all lesser species.

When considering man's triumphant journey from cave to skyscraper, one might be tempted to heave a huge sigh of relief and settle back to enjoy the benefits of humanity, whether these accrue from the reaping of a bumper crop of yams on the island of New Guinea or the excitement of a long-awaited landing on the moon.

But humanity breeds humanity. Man, blessed with the art of reason, strives to use it. Poised at the very threshold of outer space, he looks backward, too, his thoughts on the earth's problems of hunger, disease, war, overpopulation, pollution. In all the animal kingdom, only man plots to better his present and ensure his future. Only man ponders the past.

For, if man is to know where he is going, he must first learn where he has been. And, for man, life's greatest puzzle is the mystery of his own origins.

For some, the facts of man's prehistoric past are irretrievably lost in the shifting sands of time. For others, they are merely misplaced. Fossil man, his bones hidden helter-skelter in the earth, waits to be recovered by the men and women who spend their lives scratching for bits of ancient bone—the only tangible clues we have to the course of human development. This is the story of those men and women, and of the fossil men they stalk.

A book of this sort is a product of many minds and hands; it would be impossible to list here all who have contributed by discussing their theories and excavations, by offering notes and charts, and by giving support and encouragement. But special thanks must go to Larry Quade of the University of Kansas; to Dr. and Mrs. John R. Napier of the Smithsonian Institution's Primate Biology Program; to Dr. and Mrs. Alfred Johnson of the University of Kansas Museum of Natural History; to Dorothy Elliott and Irene Reynolds of Lawrence, Kansas; and to Mary Louise Birmingham and Carlotta Rosenthal of Praeger's Young Readers Department.

SHARON S. MCKERN
THOMAS W. MCKERN

Lawrence, Kansas
April, 1970

TRACKING FOSSIL MAN

The Monkey War

Legend has it that the Victorian wife of the canon of Worcester Cathedral, upon hearing Darwin's theory of man's origins, threw up her hands in disgust. "Descended from the apes!" she cried with a shudder, "We hope that it is not true, but, if it is, let us pray it may not become generally known!"

In the years since 1859, a great deal of progress has been made in the search for human origins. People no longer wince at the suggestion that they live in kinship with others of the animal kingdom. Exciting new fossil discoveries from all parts of the world point up the long history of man on earth and permit us glimpses of animal forms long extinct. Science has developed important new research tools for use in dating prehistoric bone and man-made artifacts. Dust-coated men sifting through heaps of ancient soil have laid bare many of the earth's secrets, including not one but a whole series of "missing links." Laboratory technicians, working with quick-breeding varieties of the common fruitfly, have been witnesses to evolution in action. Others, peering through high-powered microscopes only recently available, have known the wonder of unraveling—at least partially—the mysteries of human heredity.

Darwin's concept of organic evolution—the change or modification of living forms over great periods of time—has shed its unsavory reputation and grown into a respectable topic of scientific inquiry. Not in the sense feared by the saintly canon's wife, of course; for, of the four genera of great apes living today, none ever could have provided an ancestral base for man.

Darwin never claimed they could. Nor do modern scientists. What they *do* believe is that man and the apes—the gorilla, chimpanzee, gibbon, and orang-utan—shared some 10 million to 20 million years ago an animal ancestor, a generalized sort of primate that subsequently gave rise to both the modern apes *and* man. Darwin saw the ape not as man's grandfather but, rather, as a close animal cousin.

Darwin's concepts weren't new; philosophers since the time of the ancient Greeks have suggested again and again some kind of evolutionary development for existing life forms. But Darwin published his *Origins of Species* and, later, *The Descent of Man* in Victorian England, where such suggestions were blasphemous. They hit at the heart of the Biblical story of the Creation at a time when the Bible was the only permissible authority on man's origins. When Darwin dared to theorize an animal ancestry for man, he roused the fury of scientists and clergymen alike, drawing the battle lines for a great "monkey war" that would rage for more than a century.

In what must have been the classic understatement of the age, Darwin predicted that his theory, particularly with reference to man's development from a lower animal form, might prove distasteful to many. And, while evolution was an unpopular concept in Europe, it met with the most violent opposition in the United States, where some groups urged not only fines and imprisonment for those found guilty of teaching the doctrine, but also hanging and crucifixion!

American religious leaders were horrified by Darwin's revolutionary ideas. Conservative Protestants, spurred on by the purest of religious motives, paused in their fight against "demon rum" to deal with this new menace from abroad. Aimee McPherson, preaching on the West Coast, presided at torch-lit rallies where "monkey teachers" were hanged in effigy. On the Atlantic Coast, John Roach Straton waged a fiery campaign against drinking, dancing, and evolution. And, in the South, the most eloquent of all fundamentalists, William Jennings Bryan—former U.S. Secretary of State, prominent attorney, and respected journalist—joined the cause, arguing winningly for the passage by state legislatures of anti-evolution laws.

By 1920, it was a crime in numerous states to teach in any public school or college the doctrine that mankind had ascended or descended from any lower form of animal.

In 1925, a Dayton, Tennessee, high school senior asked his physics teacher about evolution. The teacher, John Thomas Scopes, recklessly explained the theory in a few brief sentences and found himself in jail for his trouble, quickly indicted for unlawfully having taught evolutionary theory.

The small, normally tranquil town of Dayton became the battle-field for the last great skirmish of the monkey war. Scopes—young, unmarried, and able to afford the loss of his job—was the ideal defendant for a test case. For the anti-evolutionists (mainly good-intentioned Christians of various denominations who believed that Darwinian theory contradicted the word of God as related in the Holy Scriptures), the Scopes trial seemed an excellent opportunity to crush, once and for all, the evolutionist forces.

They named, as commander-in-chief for the prosecution, no less fervent a champion than William Jennings Bryan, perhaps the nation's greatest orator. Pitted against Bryan and the Attorney General were the flamboyant liberal attorney Clarence Darrow and his friend Dudley Field Malone, together with defense lawyers dispatched by the American Civil Liberties Union. The issue at hand, unquestionably, was Darwin's theory of evolution. Scopes, the defendant, was all but forgotten in the heat of battle.

Throughout a long, hot July, these men argued the case under the most extraordinary circumstances yet known in American court history. Hundreds of spectators descended like locusts upon the courthouse; swarms of journalists from newspapers across the country jostled for a better view. The presiding judge, fearing that the ancient courthouse floor might collapse beneath the weight of so many trampling feet, ordered the trial moved outdoors to the courthouse lawn. Here spectators formed themselves into opposing cheering sections—the liberals vastly outnumbered —and enterprising vendors milled through the crowds peddling hot-dogs and lemonade.

Darrow, resplendent in blue shirt and purple suspenders, presented a brilliant defense, occasionally allowing his frayed temper to explode into verbal fireworks as the judge refused to admit the testimony of Darrow's expert witnesses. Once, driven to sarcasm by the heckling crowds, he turned and grumbled, "Why don't you folks cheer?" He was promptly cited for contempt and forced to apologize to the court.

Bryan, on the other hand, managed the prosecution's case with professional aplomb—at least until Darrow, unable to get his prepared scientific evidence before the court, announced that he would call Bryan as a defense witness in order to show that the Bible need not be given a literal interpretation.

Bryan endured with mounting impatience a barrage of questions regarding Jonah and the whale, Joshua and the sun, the Biblical flood, and the Genesis account of life's creation. The prosecution interrupted incessantly, trying in vain to bring Darrow's questioning back into legal bounds. Finally the exasperated Attorney General demanded to know the purpose of Darrow's relentless badgering of Bryan.

"To prevent bigots and ignoramuses from controlling the educational system of the United States!" shouted Darrow as Bryan, in a burst of fury, leaped to his feet and denounced Darrow for having cast slurs upon the Bible.

Throughout the afternoon and again the following morning, Darrow goaded his opponent, hoping to entangle him in a trap of logic. *Was Eve actually made from one of Adam's ribs? Where did Cain get his wife? What sort of whale, exactly, swallowed up Jonah—a run-of-the-mill fish, or one made by God especially for that purpose?*

Bryan, publicly humiliated and confounded by what he considered a blasphemous attack upon his religion, refused to admit defeat. Reluctantly admitting under oath that some events in Biblical history must be taken as fable, he insisted nevertheless upon a literal acceptance of other events. He retreated again and again from Darrow's questions, taking refuge in his faith.

"The Bible states it," he'd reply. "It must be so."

Queried about scientific evidence to the contrary, Bryan disavowed any interest in the concepts of geology, physiology, archaeology, or comparative religion. The Bible, he said, gave him all the information he needed to live by or die by.

Often the testimony could not be heard above the whooping and cheering crowds. And the days passed in a blur of blistering sunshine and shouted objections.

Eventually, the aged Bryan, his composure lost, seemed hopelessly distracted by the unprecedented defense. It was clearly Darrow's day. But, as wily and logical as the defense attorney was, he was still unable to circumvent the judge's prohibition against the introduction of scientific evidence.

When the battle smoke cleared, Scopes was found guilty of having unlawfully taught the doctrine of evolution and was ordered to pay a $100 fine. The verdict strengthened Tennessee's anti-evolution law and prompted a flurry of similar statutes in other states, including strong laws in Mississippi and Arkansas. But, if the defense lost the court battle, it won the war; for the well-publicized trial revived popular interest in Darwin's evolutionary concepts. Darrow's rational defense convinced the public, if not the court, that there was nothing criminal in Darwin's strange new ideas. And he had brought to the public's attention new scientific findings on earth history.

Today, only one monkey law remains on the books: that of the state of Mississippi. Tennessee's Butler Act, responsible for Scopes's indictment, was repealed in 1967. In 1968, the U.S. Supreme Court ruled unconstitutional Arkansas's law forbidding the teaching of evolution in that state. It is unlikely that Mississippi will choose in the future to enforce its anti-evolution statute.

Does all this imply that Darwin was right and the Bible wrong? Darwin's theory was advanced at a time when the mysteries of human origins were considered to be outside the realm of scientific verification; any attempt to replace the Adam-and-Eve account of Creation was viewed with hostility and alarm. Most people now regard the Bible as a work with moral import, its concepts conveyed frequently in parable, and its account of the Creation as symbolic. And so, today, we need not choose between science and religion.

Further, evolutionary theory makes no attempt to invalidate religious teachings. It does not advance a theory explaining how life began but seeks instead to discover how life *developed* through time and space. This it accomplishes by studying man as a biological organism, subject to the same natural laws as the rest of the animal kingdom, but always with reference to the fact that man is different: He can think and reason. He can remember what happened yesterday and plan for tomorrow. He has the most dexterous hands in the animal world, hands that obey with ease and agility the commands that come from his complex, highly developed brain. Man has the power of speech: He can communicate like no other animal on earth, and he is able to teach other men what he has learned, transmitting accumulated knowledge from generation to generation. In short, man has a unique capacity for culture. In all the world, only man has the ability to manipulate his own environment to suit his desires and comforts.

Not only anthropologists but scholars in almost every scientific field join to search out the hidden facts of man's present nature, and few stones are left unturned in this never-ending quest for more and better answers.

How do we come to know ourselves? Geographers trace the dispersal of mankind over the globe, tracking man as he conquers the deserts, the mountains, and the seas—and now, outer space. Sociologists and psychologists watch man as he works and plays, in the cities, on the farms, in jungle villages. Zoologists and primatologists journey to remote wilderness areas to observe and record every aspect of ape and monkey behavior, so that someday we may hypothesize the social life of our earliest ancestors. Anatomists delve inside the human form, scrutinizing each organ, however minute, to determine man's taxonomic place in nature. Serologists do the same, tracing kinships through similarities in blood chemistry. Embryologists breach even the sanctity of the human egg, watching human development before birth occurs; and geneticists peer into the cell nucleus itself, seeking the answers that eluded Darwin. In these ways, we come to know the secrets of man's present nature. And these are the simplest secrets to decipher, for they survive today to be studied at our leisure.

The past is more elusive and much more difficult to trace.

Our search begins far back in time, before the earliest of human records, beyond the physical evidence of Egypt's most ancient mummies. To discover what man was before he became human, we must track him down in the earth's surface. With any luck, we find traces of past human life: fire-blackened hearths and pits, painted cave walls, man-made tools of stone that could withstand the ravages of time. And, if we are spectacularly fortunate, we find ancient human bones.

As luck would have it, the earliest forms of man had little or no regard for the problems that would plague future historians. For the most part, they failed to bury their dead. It would have helped had more of our ancestors become fossilized, but this is no easy process. Bones left on the ground are taken and scattered—if not eaten—by animals. Even covered with earth and stone, they decay rapidly. In order to become a successful candidate for fossilization, one must die in a cave, where the bones can become impregnated with mineral salts, or fall into a lake bed, where sediment may cover and preserve the form. Recently, graves of two stone-age cave dwellers excavated in Spain by anthropologists from the University of Chicago yielded bones that were amazingly preserved. As their owners' bodies decayed, fine sediment of clay filled the cavities. The clay assumed the appearance of the original man, even to three-dimensional folds of flesh.

But such conditions aren't met very often, and there are never sufficient fossils. Only a tiny fraction of living populations die under conditions favorable for fossilization; of these, an even more minute number manage to come to light for our examination. Occasional happy discoveries, however, enable anthropologists and paleontologists to locate and preserve certain extinct forms from the distant past. With the help of modern dating techniques and sound geological principles, these are arranged in chronological sequence from ancient to modern, so that we may view developmental sequences as they occurred in nature and as they have been captured in stone. The evolutionary history of some few forms is remarkably complete. For the horse, for example, a documented sequence has been put together, dating back through the last 60 million years; we can "see" the evolution of a horse from its appearance in Eocene times as a three-toed dwarf to the proud form it bears today.

The quest for human and prehuman remains has been less

successful, and we may never find all the "missing links" that lie
hidden in the earth. Man was too clever an animal, even at the
start, to go about falling into lake beds or getting caught by cave
bears. His most critical stages of development, it is thought, occurred in forests, areas unfavorable for fossilization. Nevertheless,
sufficient fossils have been uncovered for us to trace a general
development for man on the basis of the present evidence. And we
can help to fill in the gaps with evidence of different sorts. Often,
even when the bones are not preserved, the teeth—as the most
durable of all body parts—survive to hint at form and function.
From them, we can deduce species, whether human or nonhuman,
and size. Often, we can tell which sex is represented. Almost always we can infer dietary habits. In addition, archaeologists uncover crude stone tools which have much to say about the level of
culture and intelligence attained by their lost makers. From
geologists, we learn of climatic changes in the earth's history.
From botanists, working with preserved pollen, we learn the
climate of a given region at a specified point in time—all of which
will be discussed in a later chapter.

The man who started it all made no use whatever of fossil evidence, although human fossil material was beginning to accumulate in museums all across Europe in his time. Darwin lacked the
research tools available to us today. He had no knowledge of the
laws of genetics and, seemingly, little interest in the fossils that
would validate his theory.

What he did have was an observant eye. He had noticed, particularly on his ocean voyages to the Galapagos Islands, that there is a
tendency for all organisms to multiply, so that more individuals
are born in each generation than can possibly survive to adult-
hood, a phenomenon earlier recorded by Thomas Robert Malthus.
At the same time, he noticed that the numbers of any given
animal species seemed to remain more or less constant. And he was
struck by the amazing diversity of life forms; all living things
vary. Further, offspring resemble their parents but do not duplicate them exactly.

From these three primary observations, Darwin made two
important deductions. First, he perceived that, since there are
more new individuals born into each generation than the environment can support, there must occur a struggle for survival among

the world's living organisms, a constant scramble on the part of each individual to take for itself sufficient food and space with which to maintain life. Darwin called this phenomenon the "survival of the fittest." But the most fit need not always be the strongest or the most ferocious. It is not the biggest cat that takes a lion's share of available food supplies, but the fastest, the best able to capture game, or the most able to sniff it out and track it down. Speed is less important to the giant tortoise; for him, it is the durability of his shell that protects him. The chameleon, able to change color so as to blend with its surroundings and become almost invisible to the eyes of lurking predators, has a better chance of escaping the notice of its enemies than does the common lizard, who depends more on speed and agility.

Secondly, Darwin perceived that only a fraction of the infants born live to become reproducing adults: the most fit fraction. Those individuals with some advantage have the best chance for survival and thus for reproducing their own kind. Organisms that survive and multiply with ease are those that have successfully adapted to their environment. And adaptability, in nature as well as in human life, is of prime importance; it implies the ability of an animal to cope with climatic changes by edging into new ecological niches, to find new foods when old foods are not available, to catch the eye of a prospective mate and attract him.

Well-adapted individuals live to mate and reproduce, yielding offspring that resemble them in whatever advantageous characteristics ensured their survival in the first place. The less fit—the weaker, the slower, the less fleet of foot—are eliminated; the sick and defective die out. This is *natural selection,* the force that Darwin deduced must operate to carry the processes of evolution into action.

Nature abounds with dramatic examples of adaptation. Desert plants store water so as to survive the most arid of droughts. Others look like rocks and so escape the alert eye of a would-be grazer. Birds like the peacock develop magnificent plumage with which to attract mates. Mammals like the whale and the bat adapt to exploit new environments. Modern scientists know the value of successful adaptations. But they have modified Darwin's theory to reflect a new realization that it is *fertility* that most crucially determines survival. The form able to reproduce itself in huge numbers is the most likely to endure through great periods of time. This is why man and his mammalian relatives are so badly outnumbered by the fish and insects, which reproduce in great quantity. Compared with the 3,200 known species of living mammals, there are 20,000 species of fish and more than 800,000 species of insects! Man, with his habit of single births, is at a disadvantage. But he gains through the fact that he knows no mating season: *Homo sapiens* is able to reproduce the year 'round.

However crude and unproved Darwin's theory, it succeeded in demonstrating the amazing diversity of plant and animal forms across the world, and helped to explain the similarities and differences apparent among them. It showed, too, the kinship existing between even the most dissimilar life forms and demonstrated that all life springs from a common beginning. Darwin's essential contribution lay in his recognition of a mechanism through which evolution, or change, appeared to be taking place. That mechanism is natural selection.

From this modest beginning, nineteenth- and twentieth-century scientists built a new and more comprehensive hypothesis of life's development. They have identified the raw material on which natural selection operates to produce change; that is, *mutation,* a spontaneous change in the gene that is transmitted to future generations. With the help of constantly refined research tools, they have defined the mechanics of heredity, coming to learn the laws of genetics and to understand the processes of change. They have transformed the "theory of evolution" into "evolutionary theory," a rational statement of all that is known concerning the development of species—human and nonhuman. And, with the accumulation of new fossil discoveries, they are piecing together, with fewer and fewer gaps, a complete history for man from his beginnings to the present. But these scientists—however dedicated, however successful in providing the proof that Darwin lacked—aren't the principal characters in our story. Our starring actors are those men and pre-men and man-apes who peopled our world in the past. This is *their* book.

2

The Ancient World

If we are to build a complete life history for man, we shall have to search for clues in the darkest past, beginning not with man's immediate primate ancestors but with the preserved remains of the earth's most remote life forms. The best evidence for man's development comes from collected bits and pieces of fossilized bone, and the same is true for the plant and animal forms that preceded man on earth.

Fossils provide the language in which the saga of evolutionary progress is written. Like human languages, fossils come in various forms. Some are *petrifactions,* produced when the original constituents of bone disintegrate and are replaced by minerals that duplicate the shape and structure of the lost bone. The human fossils from Spain, mentioned in the last chapter, are petrifactions. In other cases, a bone may be covered with some fine material—silt, perhaps, or volcanic ash—that later hardens. When the bone disintegrates, it leaves behind a natural mold into which the anthropologist need only pour plaster or plastic compound in order to make a cast. At the ancient village of Pompeii, archaeologists recovered in this fashion more than fifty full casts of human bodies, all representing the victims of a fiery volcanic eruption that blanketed the entire town with ash two thousand years ago.

Fossilization takes place only under special conditions, most often in shallow water where a fallen body might rapidly be covered with silt or fine sediment. Hence, petrifactions and molds of land animals are extremely difficult to come by. Even when recovered, they usually reflect only the hard parts of an animal body, enabling us to reconstruct size and shape but leaving us to wonder about such features as skin, hair, or color. Every once in a while, however, fortune smiles upon the determined fossil-hunter: extraordinary accidents of preservation yield up whole bodies complete with soft inner parts. This happens occasionally when an animal is suddenly frozen (as in the case of woolly mammoths found encased in solid blocks of ice on the snowy ranges of Alaska and Siberia) or when death occurs in pits of tar, peat, or asphalt. When a bone or body falls into natural peat bogs (deposits of moist vegetation accumulated in packed layers), a chemical reaction occurs to prevent decay. Natural tar pits in California, Ecuador, and elsewhere have preserved for us the remains of giant condors, saber-toothed tigers, and other hapless victims of the sticky tar. Pools of natural asphalt made similar traps, saving for us other forms from the past. But perhaps the most spectacular discoveries are those that occasionally appear in amber. Many luckless insects landed in the resin that exudes from trees; these were caught and preserved as, over many hundreds of years, the resin hardened into amber. Finds like these provide rare opportunities for studying extinct forms never before seen by human eyes.

More frequently, scientists must make do with the more common fragmentary fossils or tracings and impressions left on soft moist surfaces that gradually hardened into rock. But even these meager clues help science to reap a rich harvest. From them, we deduce the forms of plants and animals that thrived millions of years ago. Our museums are filled with such fossil treasures: stone etchings of delicate prehistoric ferns pressed upon ancient silt, plodding dinosaur tracks left beside a muddy creek-bed, graceful tracings of primitive starfish that once drifted over the ocean's floor in search of food.

Fossils were not always held in such high esteem. Greek scientists before the time of Christ who came upon fossils correctly but offhandedly identified them as the petrified remains of prehistoric animals. Although numerous Greek scientists hypothesized an evolutionary development of one sort or another for the earth's animal forms, they attached no great significance to the fossils they found. Aristotle, one of the most influential of the early Greeks, believed that fossils originated in the rocks, did not evolve, and therefore told no history. During the Dark Ages, a Persian scholar named Avicenna re-examined Aristotle's writings, compared them with his own observations, and came to the conclusion that fossils were produced, like sculptures, by the earth's natural movements—and that many of them were deliberately formed for the sole purpose of fooling the men who sought to study them.

About A.D. 1500, the great Leonardo da Vinci turned his attention to the marine fossils he found while building canals high above sea level in Italy. It seemed clear to him that the biblical deluge could not have carried so far inland the remains of so many creatures from the sea. He deduced that thousands of years earlier, the seas must have covered all of Italy; as the waters receded, dead animals fell upon the ground, their remains hardening later into stone. Da Vinci, hailed as a genius for his achievements in fields that ranged from art to anatomy to architecture, was severely criticized for his views. Aristotle's explanation of fossil-laden rocks had been accepted by the Church as an official teaching; the formulation of any theory to the contrary was construed as a dangerous act of blasphemy. People so feared the judgment of an angry god and the condemnation of their clergy (not to mention the wrath of skeptical friends and neighbors) that they began to look upon all scientists with suspicion. Da Vinci fared well enough despite his eccentric opinions regarding fossils, but later scientists would bear the brunt of vigorous persecution. Giordano Bruno, a luckless Italian philosopher who publicly denounced Aristotle's ideas, paid the ultimate price for his recklessness; he was driven from his home, imprisoned for seven years, excommunicated from the Church, and finally burned at the stake.

Understandably, most scientists fell silent on the subject of fossils. The public began to manufacture their own interpretations of the lifelike forms etched in rock. Soon tales of great dead giants, dragons, demons, and crystallized monsters swept across Europe. So long as one did not speak out against the dogma of the Church, the collection of "dragon" bones was considered a harmless enough pursuit; monster bones began to turn up almost everywhere. Many of these were combined with the bones of unrelated animals and "reconstructed" into complete skeletons, resulting in a grotesque collection of bizarre and freakish creatures.

As early collectors scoured Europe's countryside in search of dragon bones, some excellent fossil material began to accumulate. Unfortunately, these first fossil-hunters had no way of predicting the importance of geology in evaluating fossil material. Bones were plucked from the ground and stored in private museums without the slightest notation of place, depth, or circumstance of discovery.

A fossil whose age cannot be determined is of little scientific value. It becomes a mere museum curiosity, interesting enough on the shelf but stripped of the clues that place it in its proper context. Today every effort is made to deduce as accurately as possible the age of each fossil discovery. Scientists accomplish this in the same way that geologists date rocks. On land or in water, geologic deposits are laid down in horizontal layers, one after the other. Each layer or bed is older than the layers above it, and younger than the beds below. One might reasonably expect, on the basis of this single basic principle, that in undisturbed stratigraphy any fossil excavated from a given layer must be older than fossils found in layers above and younger than fossils found in layers below. And if the task of dating skeletal remains were indeed so simply dispatched, anthropologists could spend their summers lolling around their neighborhood swimming pools rather than hunching bleary-eyed over cluttered laboratory tables.

Frustrations abound when geologic deposits are disturbed through erosion, volcanic action, sliding, faulting, or other natural earth movement. Then, to the dismay of men who search for missing links, fossils begin to wander into inappropriate layers. Intrusive burials, occurring when fossils make their way into levels in which they do not belong, must be noted and dealt with, usually through the application of trace-element tests. One such test is fluorine analysis, used to estimate the *relative* ages of bones. Scientists have discovered that the chemical composition of buried bone changes through time as the bone absorbs tiny quantities of fluorine from the surrounding soil. Naturally, the older the bone, the more fluorine it will have absorbed and thus will contain at the time of testing. The fluorine test was put to good use in solving a sixty-year-old fossil controversy centering about a mass of human bones known collectively as the Galley Hill skeleton. In 1888, British diggers recovered portions of a human skeleton (together with stone tools) in gravel deposits that were found to be 200,000 years old. The tools were crudely fashioned, as if made by primitive stone-age men. But the skeleton itself differed not at all from that of modern man. The implication was that people who looked exactly as we do today were roaming about England in the middle of the Pleistocene, a suggestion that made hash out of existing theories of human evolution and set many an anthropologist to gnashing his teeth in his sleep. The experts immediately took sides, one camp maintaining that such a modern-looking skeleton simply could not be 200,000 years old, the other doggedly insisting that the geologic age of the deposit from which the skeleton was taken was unquestionable. Finally, in 1948, scientists subjected the Galley Hill skeleton to the newly developed fluorine analysis. The fluorine content of the human bones was measured and compared with that of mid-Pleistocene mammal bones found at the same site. Results proved the skeleton to be intrusive, a recent burial that had worked its way down into earlier deposits. When further tests supported this finding, a huge sigh of relief went up from those who had pondered for decades this enigmatic fossil man from Galley Hill.

While fluorine analysis works to determine the relative ages of bone, other methods measure age in terms of actual years, plus or minus a few centuries so as to allow for error. The best-known test of this type is the carbon-14 method, a technique particularly valuable because it can be applied not only to bone but also to other organic materials—seeds, shell, wood, charcoal, and the like.

Thus carbon-14 is often used to help date habitation sites which yield no human bones.

Carbon-14 is a radioactive isotope present in all living organisms. So long as life continues, this element is absorbed into the body at a steady rate. It also diminishes, at a rate equal to absorption. When death comes, however, no new amounts of carbon-14 are taken in, while disintegration of carbon-14 atoms continues at a steady rate. The carbon-14 isotope has a half-life of 5,760 years; that is, the rate of disintegration (originally 15.6 disintegrations per minute per gram of carbon) is reduced by one-half each 5,760 years. Due to the stability of the decay rate, scientists can calculate the date at which an organism died by measuring the amount of carbon remaining in a sample of bone or other organic matter.

Sometimes a sample is *contaminated,* as where there occurs an accidental decrease or increase of carbon-14 following the time of death. Early atomic research tests conducted in the state of Nevada so significantly raised radiation levels that samples being processed as far away as Chicago were rendered worthless. The same problem arises with other radioactive dating methods, including the usually reliable potassium-argon test as well as those utilizing the transformation of thorium to lead, uranium to lead, and rubidium to strontium.

Ideally, archaeological samples are protected from the very instant of discovery, sealed immediately in lead containers and rushed to process in specially constructed laboratories. But such protective measures are seldom practical; the hard-working archaeologist, habitually pressed for both time and money, does the best he can by quickly wrapping his samples in aluminum foil and encasing these in sealed glass jars. Whatever the limitations, radioactive dating techniques provide the best tools yet devised for determining the age of geological and archaeological specimens.

The need for such complex and sophisticated methods emphasizes the fact that the time has passed when an archaeologist could locate, excavate, and report a site alone. Not only must he depend upon geochemists and geophysicists for conducting extensive laboratory tests, but he must also draw upon the services of a wide variety of other specialists. Before he dares pick up his spade, he must consult geologists and geographers in order to satisfy himself that he has chosen a likely spot. The excavation of a fossil bed is a lengthy, expensive, and tedious process; before beginning, the archaeologist must have some reason to believe that he will find what he's looking for.

Once digging begins, an army of experts descends upon the site. Trained artists and photographers record each step of the excavation so that the procedure may be reconstructed later on paper. During field work, the supervising anthropologist or archaeologist summons specialists as needed. Palynologists come to collect tiny samples of fossilized plant pollen, hoping to reveal details of climate at the time the site was inhabited. It was from just such studies that we learned of Neanderthal man's funeral rites; pollen taken from grave burials indicates that Neanderthal man buried his fallen kin beneath a blanket of floral offerings.

Paleontologists examine all recovered animal remains, deducing from them the total ecology of the region and the eating habits of fossil man. If dating tests are inconclusive, associated animal remains may help to pinpoint geologic time. Soil experts remove cores of earth to shed light on ancient vegetation patterns. If the site is a young one (no older than several thousands of years), two-part varves may be taken from sedimentary rock to yield valuable information on past seasons. During summer, melting glaciers deposit great quantities of sediment on lake bottoms (thus summers are represented on the varve by wide, heavy bands). In winter, water ceases to flow, resulting in scant deposits (represented on the varve by narrow bands). Similarly, wood beams recovered from recent habitation sites provide a record of climate as well as age. The width and spacing of tree-rings result from the combined effects of light, moisture, and temperature. In arid regions (as in the American southwest) where wood is frequently preserved, archaeological sites less than a thousand years old can be dated reliably from such tree-ring dating. The technique is called *dendrochronology*, a ten-dollar word that means, quite simply, tree-ring analysis.

In older sites, particularly those dating from stone-age times, petrologists are asked to identify the materials used to make stone tools and weapons. If artifacts are found to be made from types of stone not native to the region under examination, investigators locate the nearest source of the stone and trace ancient migration routes. Archaeologists working near Tres Zapotes in Central America, for instance, came upon colossal carved heads made of stone available only at Mount Tuxtla, ten miles distant. Each carved head weighed at least ten tons. The means by which prehistoric peoples, lacking either wheel or domesticated pack animals, transported these gigantic stone blocks over ten miles of rough terrain remains a mystery. But archaeologists have located the routes taken by those long-vanished Indians in obtaining these blocks and, in the process, made new, unexpected discoveries at additional sites along the way.

Human bones taken from ancient burials go directly to the physical anthropologist, a specialist whose job it is to evaluate the physical characteristics of any skeletons recovered from the site. Through a detailed study of each skeleton, he can determine sex, height, approximate weight, racial kinships, and age at the time of death. Often, he will find in the bones ancient evidence of disease or specific injuries suffered during the individual's lifetime; sometimes he can pinpoint the actual cause of death. Working with statisticians and taxonomists, he lists all physical traits noted in his analysis of the skeletal material and tries to fit this particular fossil population into the current evolutionary scheme. This is how he is able to trace man's development, tracking down a fossil man's ancestors and his decendants.

Even field laborers must be trained to search out and preserve the most minute traces of ancient man's existence. Shovels abound at any digging site, but these are wielded gently so as not to destroy bones or artifacts that rest hidden in the earth. Most often it is with toothbrushes and whisk-brooms, on hands and knees, that the real work of excavation is done. Earth, rock, and pebbly soil are removed inch by careful inch; all are passed through a mesh screen before discard so that tiny bits of bone or fragmented artifacts inadvertently overlooked in digging are not irretrievably lost. When bones or implements are found, they are numbered, photographed, and catalogued before removal. Fragile bones must be treated with hardening agents (epoxy or plaster) to keep them from crumbling upon sudden exposure to air.

No site is abandoned simply because it fails to reveal human fossil or cultural evidence. In order to learn man's complete story, we must begin as far back in prehistory as possible, linking in chronological sequence the fossil remains of both plants and animals to illuminate all of life's history up to the present time.

No one knows how life began on earth. Many scientists of the past, earnest in their other investigations but overwhelmed by the immense mystery of life's origins, washed their hands of the entire matter by declaring that life *always* has existed, a fact ably disproved by modern geologists. Others favored the more exciting outer-space theory, an adventure in conjecture that suggests the tantalizing possibility that archaic life forms arrived on earth in clouds of cosmic dust at some dark, distant point in time. Such a theory conjures up tempting visions of extraterrestrial life, but it fails to explain how tiny, unprotected creatures, however hardy, managed on their way in to survive the earth's deadly radiation belts or, indeed, how life arose elsewhere in the first place. For the scientists of Darwin's time, this matter of life's origin was a *spiritual* rather than a *scientific* concern. Relying upon a literal interpretation of the Bible, they deduced that each species of plant and animal life had been separately and divinely created. Scientific investigations centering about the workings of natural laws could serve only to contradict the word of God. Science stood in direct conflict with religion.

No such conflict exists today, for science and religion are increasingly tolerant of one another. For those who acknowledge the existence of a supreme being, science works to reveal the splendor of life's infinite mysteries and so reinforces faith in God's powers. For their part, scientists are not in the business of disproving the operations of supernatural agents, nor do they attempt to do so in their spare time. Science neither ponders man's reason for being nor argues whether natural laws were formulated by divine command.

Scientists do, however, offer alternate theories of creation based upon the natural laws they observe. We have said that evolutionary theory makes no statement regarding life's *origins*. And it is true that anthropologists are concerned instead with the ways in which life *developed* after its origin—for all present and future life forms develop from pre-existing forms. Since anthropologists seek to uncover traces of man's ancestors, however, they are interested in the findings of other scientists regarding the processes by which life probably began on earth. The life-origin theory most commonly accepted today involves *spontaneous generation,* a phenomenon that occurs only in the absence of oxygen. This process cannot occur naturally today (although scientists can simulate it under controlled laboratory conditions), because once life was established here on earth, the atmosphere began to contain oxygen.

It is believed that high temperatures prevailed after the formation of the earth. Water was present only in the form of vapor or steam. No free oxygen existed, although simple chemical compounds—probably containing carbon and nitrogen—were abundant. With the cooling of the earth, liquid water was formed. Then, at some unknown instant in time more than three billion years ago and under ideal conditions of light, moisture, and moderate temperatures, certain simple compounds came together in chance association to result in the production of the first substance capable of reproducing itself: living matter.

And so life began.

Spontaneous generation (if this is indeed the process by which life originated on earth) took place in or near the water. Tiny one-celled microorganisms began to drift about in the coastal waters. And they multiplied—one cell split into two, or two into four, or four into eight—and so it went until the waters teemed with life.

But the fate of many of those tiny, primitive creatures was starvation: they had no means of obtaining food. Some of the first cells developed green matter that trapped rays of sunlight to yield energy; with this force, these cells—our first plants—took gas from the air and water from their surrounding environment and manufactured their own food. Cells not capable of photosynthesis fed upon those which were. The division between plant and animal was achieved in those most remote days of prehistory.

For men who hunt for traces of ancient life, the search for fossils representing these earliest one-celled microorganisms holds more drama and excitement than any other scientific pursuit. And few quests prove to be so frustrating, for our fossil record of life in its earliest stages is meager and scant. The animals and plants of that distant era, not nearly so abundant as in later times, were generally soft-bodied, unlikely to leave evidence of their passing.

The few fossils deposited in pre-Cambrian times were almost surely doomed to destruction as the rocks in which they were contained *metamorphosed,* or were altered by heat and pressure. And the fraction of fossils that remain to us, hidden in the depths of the earth's surface, are difficult to find.

There is no question, however, that life did exist in pre-Cambrian times. Our earliest evidence comes from deposits of graphite, the material used as "lead" in our pencils. Graphite is composed of carbon, which in turn represents the decomposed remains of ancient vegetation. There can be no doubt that there was an abundance of living organisms available during the Archeozoic era to form these extensive deposits of carbon.

The oldest true fossils yet known come from Swaziland, Africa, where scientists have discovered the remains of microscopic spherical-shaped organisms measuring no more than 0.000008 inches in diameter. These micro-fossils, containing a mass of blue fluorescence, were subjected to extensive chemical analysis and found to contain three different amino acids, the basic compounds of protein. The tiny fossils were taken from stone deposits believed to be 3.2 billion years old.

As primitive as the Swaziland micro-fossils are, they do not represent the earliest, most primitive one-celled organisms to appear on earth, and it is unlikely that we ever shall recover samples of those first forms of life. There are great gaps too in the later chapters of fossil history and huge chunks of time for which we have no fossils at all. These gaps may never be filled; many steps along man's evolutionary road must remain obscured by time's mists forever. Nevertheless, our present accumulation of fossils is sufficient to permit a linking of major animal forms, and we can extrapolate from one to the other. In this way, we can sketch a faint and cautious outline of life's development from the earliest geologic eras.

But how does one visualize geologic time? The human mind, accustomed to dating events in terms of weeks and months and years, boggles at the thought of a *billion* years—or even a million. Such enormous segments of time are simply incomprehensible to the cleverest of us. Fortunately, some twenty years ago a man named James C. Rettie took the trouble to whittle down these figures to a manageable size. He imagined a camera hung in outer

Geologic Time Scale

Era	Period	Epoch	(Years Before Present)	New Life Forms
CENOZOIC	Quaternary	Recent	0–8,000	
		Pleistocene	8,000–2,000,000	Man
	Tertiary	Pliocene	2,000,000–12,000,000	Proto-hominids
		Miocene	12,000,000–28,000,000	Pongid radiation
		Oligocene	28,000,000–40,000,000	Pongids
		Eocene	40,000,000–60,000,000	Cercopithecoids
		Paleocene	60,000,000–70,000,000	Prosimians
MESOZOIC	Cretaceous		70,000,000–130,000,000	Placental mammals
	Jurassic		130,000,000–170,000,000	Pre-marsupial mammals
	Triassic		170,000,000–200,000,000	Archaic mammals
PALEOZOIC	Permian		200,000,000–250,000,000	Amphibian decline
	Carboniferous		250,000,000–300,000,000	First reptiles
	Devonian		300,000,000–350,000,000	First amphibians
	Silurian		350,000,000–400,000,000	Land plants
	Ordovician		400,000,000–500,000,000	Ostracoderms
	Cambrian		500,000,000–750,000,000	Marine invertebrates
PROTEROZOIC			750,000,000–1,000,000,000	Primitive marine invertebrates
ARCHEOZOIC			1,000,000,000–3,000,000,000	Unicellular life

space, trained upon the earth by inhabitants from another planet. The hypothetical camera, fitted with super-telephoto lens and rigged for time-lapse photography, would snap one picture per year for every year of earth history. The resulting film, however imaginary, is an epic production of life's development. No mati-nee feature this, the movie—run continuously on a motion-picture projector at normal speed (twenty-four frames per second)—would provide a full year's viewing. Let's begin the film at 12:01 A.M., January 1st, and sample its highlights.

Through all of January, February, and March, we see no ac-tion, not the slightest stirring of life on earth. But things pick up about the first of April as our single-celled organisms make their debut. Within a week, many of them begin to group together to form colonies, a few learning to utilize calcium derived from sea water to build protective shells about their bodies; colonial algae produce great prehistoric reefs as much as twenty-five feet long, and equally wide. April's life forms are sedentary creatures, un-able to move about in search of nourishment. Fastened to the ocean floor, they survive by sifting through silt and sea water for the tiny bits of food these contain. Immobile, they are easy prey for larger organisms that seek to devour them.

For in Proterozoic times (the middle of April in our film), some few archaic sea creatures learn the secret of locomotion, and a new, more competitive way of life emerges. Even limited move-ment provides a means for escaping predators and permits the exploitation of new food sources. Our animated actors multiply and spread. Now we begin to see diverse forms of jellyfish, sponges, starfish, sea worms, and sea urchins. These are soon joined by the curious *nautiloids,* small marine animals that look much like our modern squids but appear to carry snail-shells on their backs. The *trilobites* reign supreme; bizarre predators with jointed legs and segmented bodies, these are our first representatives of the arthro-pod group, that taxonomic category in which all our modern crustaceans, spiders, and insects belong. By the end of April, the trilobites rule the seas as the dominant form of life.

With the beginning of Paleozoic times (about the first of May), warm ocean currents from the south scatter life forms along the marine seaways and we note an increasing diversification of sea creatures. *Brachiopods* appear, soft-bodied creatures with upper and lower shells and armlike parts that extend from their mouths. Sea scorpions become plentiful; some of their descendants will grow to a frightening length of some six feet.

So far, we've seen nothing but invertebrates, animals without backbones. The armor exhibited on the bodies of snails, insects, starfish, and the like is not bone but an organic substance deposited externally. But by Ordovician times—toward the end of May in our imaginary movie—we encounter the *ostracoderms,* first of the vertebrates and thus man's most distant relative. First to possess a living, growing skeleton, the ostracoderm is a primitive jawless fish. Plates of bone formed in the skin cover his head as well as his body. Inside, there is an internal skeleton, probably constructed of cartilage. The inner skeleton lends stability, the outer one protection against predators.

Certainly the addition of a skeleton is a vast improvement over more primitive life forms. But we can see that jawless feeding is both inefficient and restrictive; without jaws, the ostracoderm is forced to remain near the bottom of the sea, sifting silt and sediment for food. One or more varieties of ostracoderms eventually evolve jaws, enabling the resulting forms—called *placoderms*—to become active predators. Boasting jaws, they can pursue prey and, freed from the ocean's floor, they begin to explore vast new bodies of water, both fresh and salt.

Our hypothetical movie makes it obvious that evolution is not an orderly process, nor a predictable one. Evolution proceeds in spurts and dashes, its meandering route mapped largely by chance combinations of body structure and environmental circumstance. We can't tell why evolution occurs or what triggers its occasionally sudden acceleration, and we won't have these answers even at the end of the film. But we've already learned to recognize certain patterns that seem to occur over and over again in the evolutionary process. One of these is *adaptation,* the ability of a species to change its structure, form, or function through time so as to better adjust to its environment and enhance its chances for survival. It is through adaptation that certain varieties of tough-skinned desert plants will later develop the capacity for storing rain water against prolonged drought, and that numerous varieties of moths will evolve protective coloration so as to blend into the background and become almost invisible. The lifespan of any single plant or animal, of course, is too brief to permit such accommodations. But our film illuminates again and again the gradual changes that take place in a species as a whole as it adapts through time. In the case of the moths, those varieties that escape the notice of enemies will endure, reproduce in great numbers, and pass on to their descendants the defensive traits that favor survival—in this instance, protective coloration.

Another pattern is *radiation* (not to be confused with the term commonly used in physics). Evolutionary radiation involves the development of vast new varieties of life forms from a single ancestral stock. When a new, successful form appears on earth, its descendants often can take advantage of ecological niches previously unexploited. Thus new forms frequently evolve rapidly to produce great numbers of diverse types. For instance, *Tomarctus,* a wild wolflike mammal that lived 15 million years ago, gave rise to the hundreds of different breeds of domesticated dogs that we see today. Whether he chooses the huge, hardworking Saint Bernard or the lap-sized, more decorative Yorkshire Terrier, man deliberately breeds dogs to suit his own purposes and pleasures. The same process that permits such wide diversification today under human supervision operated naturally in the remote past, as we shall see as we return to our hypothetical film.

It is now June, late Ordovician times. The placoderms, highly successful predators, have evolved in all directions. Some have become cartilaginous fishes; we begin to recognize in these the probable ancestors of modern sharks, rays, and skates. Another group has given rise to a vast number of bony fishes, the forerunners of most of the 20,000 different species—ranging from aquarium guppie to seafaring tuna—that live in the present. Yet another group evolved in quite another direction: they developed lungs!

Most fish take oxygen from the water, gulping down the liquid through the mouth so that it travels down the throat and exits the body through the gill slits. These fish cannot live out of water; their survival depends upon an abundant supply of fresh water. Had we been watching our film closely, we'd have noticed that throughout the month of June the earth has been ravaged by frequent and severe droughts. Fish venturing into inland waters were trapped in dried river beds and perished. The lobe-finned fishes, however, descendants of one of the many specialized placoderm branches, were able to survive. And what a lucky thing this was for us, for some of these extraordinary lunged fish would evolve, in time, into the world's first land vertebrates, the amphibians.

By the middle of the Silurian period (July in our imaginary movie), the first land plants begin to dot the landscape—a potential source of food for the earliest land creatures. And what likelier candidate for this important role than our curious lunged fish? The lobe-fins, more properly called *crossopterygians,* not only boasted primitive lungs but also the barest beginnings of legs; the stubby fins that give them a nickname contained muscles and bones, anatomical equipment that would be invaluable during the transition from water to dry land. Some of the lobe-fins, caught in drying ponds during the seasonal droughts, began to stump their way across land in search of other pools of water. A few hardy varieties, finding on the land a new abundance of food and a total absence of enemies, remained to evolve into full-fledged amphibians.

But it is no easy thing to be an amphibian. Despite its success in pioneering the first great land invasion, the amphibian's ties to the water are strong. Eggs must be laid there; there they hatch quickly in an underdeveloped state to fend for themselves until mature. And in the water newborn amphibians face fierce competition from the fishes. Most of the earliest amphibians are doomed to quick extinction. Only the frogs, salamanders, and *caelicians* will survive until modern times.

Nevertheless, some of the earliest amphibian forms persisted in the absence of land enemies long enough to trigger a large-scale radiation of descendants. And by Carboniferous times (September in our film), we witness a domination of the earth by vast numbers of diverse reptiles. Totally emancipated from the water, reptiles exhibit strong, improved limbs, more efficient shoulder and pelvic girdles, and teeth that automatically replace themselves when lost. All of these new features favor full conquest of dry land. But the single most important reptilian innovation is its mode of reproduction. The reptile egg is large and covered with protective shell; no longer must it be laid in water to prevent drying. While amphibian young are hatched in water and forced to fend for themselves until fully developed, reptilian young remain in the egg, nourished by a plentiful supply of yolk, until the same period of development is achieved. Only then must the young reptile venture out into a harsh and competitive world. Protected from predation through their most vulnerable stage of life, the reptiles survived in great numbers.

From the ancestral *Seymouria,* an intermediate form seemingly half-reptile, half-amphibian, the reptiles evolved and flourished, giving rise to vast numbers of exotic specializations. Huge herbivores, or plant-eaters, reached gigantic proportions; *Brontosaurus* is estimated to have weighed some thirty tons. Giant carnivores, or meat-eaters, ruled the land, feeding upon weaker prey. *Tyrannosaurus,* fifty feet long and twenty feet tall, is believed to be the largest meat-eater ever to stalk the earth. Several groups, including the *plesiosaurs* and *ichthyosaurs* returned to the water. Another group, the *pterosaurs* or *pterodactyls,* became adapted for flight and invaded the air. By the end of the Permian period, the reptiles are supreme. If the Paleozoic was the "Age of Fishes," the Mesozoic is indeed the "Age of Reptiles."

Our film is so filled with the activity of these huge, successful reptilian beasts that we almost fail to detect the presence of other, smaller forms. But they are there. Primitive birds appear, some with teeth, all strange feathered creatures with warm blood and hollow bones. Making their appearance too are the first mammals, rat-sized creatures so small and timid that they almost escape our notice.

The huge reptiles have dominated center stage throughout September and October of our hypothetical time-lapse movie, becoming more perfectly adapted to their moist, swampy environment with every passing day. They have achieved a comfortable adjustment to their surroundings, developing along the lines that best suit the atmosphere in which they live. In short, they have become highly specialized creatures; they thrive as no other animals so far in earth history. Evolution favors specialization; because greater adaptation betters the chances for survival, those animals most highly adapted survive in greater numbers, becoming more specialized with each new generation.

But there is a terrible price to be paid for over-specialization.

Any animal committed to a single way of life may flourish only so long as conditions remain unchanged and that way of life endures.

Toward the end of the Mesozoic era—about November—comes a gradual lowering of the earth's temperatures. The immense swamps and lush vegetation begin to recede. Soon the giant herbivores cannot obtain the vast food supplies needed to sustain themselves. They are destined for extinction, together with the carnivores that feed upon them. By the end of November, the dinosaurs have vanished, leaving behind them only a few crocodiles, snakes, turtles, birds, and primitive mammals. The Age of Reptiles is over.

During Tertiary times—the first part of December in our film— conditions were ideal for the rise and spread of mammalian forms. Rid of the domineering reptilian monsters at last, the earliest mammals found food supplies abundant and space almost unlimited.

Most mammals are generalized animals, able to take advantage of diverse environments in order to survive. And there are obvious other advantages to being a mammal. First, its warm blood helps to maintain a stable body temperature and enables the mammal to lead a more active life than did its sluggish, slow-moving reptilian predecessors. Secondly, the mammalian skeleton is strong and well-developed, with limbs more efficiently arranged and a differentiated vertebral column, its separate elements varied in shape and size according to their location along the length of the spine. Dentition too is differentiated. While reptilian teeth are all the same and restricted to the grasping or tearing function, mammalian teeth are specialized so as to serve a number of different functions. Poke your tongue about your own teeth and see how the front (incisor) teeth are designed to cut and shear, the canines to tear, the molars and pre-molars to hasten the efficient grinding of food.

And there are still more advantages. The mammal experiences *viviparous* birth; that is, he develops inside the mother's body and then is born live, unlike the reptile which must develop in a shell before hatching. While mammals produce fewer offspring than reptiles do, the capacity for live birth (coupled with an extended period of parental association and protection) assures survival in greater numbers. Mammals have mammary glands, too, to nourish the young after birth. And the mammalian brain is characterized by an expansion of the *cerebrum,* that critical portion of the brain that will later distinguish man from all other animals. From the first mammals, through the prehistoric primates, we inherit this developed cerebrum, the organ which functions to provide the capacity for memory, intelligence, and emotion that makes us human.

Among the most primitive mammals are the *prototherians,* strange creatures that possess all the requisite mammalian characteristics but one: their young are encased in eggshell, like a reptile. Of these, only the spiny ant-eater and the incredible duck-billed platypus survive to modern times. The Jurassic period ushers in the first *metatherians,* or marsupials—the pouched mammals. They bear their young live but in an underdeveloped state, then protect them in the pouch until they are sufficiently mature to face the hazards of the outside world. Modern marsupials include the kangaroo, koala, and wombat of Australia, and the American opossum.

But it is the *eutherian* group, appearing during Cretaceous times, that interests us most, for this is the ancestral stock that will spawn the first representatives of man. Eutherian mammals are characterized by the possession of a *placenta,* that miraculous membrane made up of a combination of maternal and embryonic tissue that nourishes the unborn young while housed in the mother's body. The placenta provides both oxygen and food from the mother's blood and carries away body wastes discarded by the growing embryo. Such a method of reproduction gives the embryo sufficient time to develop the extraordinary organs—including a complex brain—that distinguish the adult. Eutherian or placental mammals quickly outnumbered the other groups early in mammalian history, for their offspring could develop slowly and safely in the protection of the mother's womb. And most of our modern mammals are placental, sufficient proof of the efficiency of this method of reproduction.

Last January, we settled back to scan an imaginary time-lapse movie that runs more than 2,000 times as long as *Gone with the*

Wind. Here it is, nearly Christmas, and we've yet to glimpse the hero of our tale. Man isn't scheduled to appear until noon on the final day of the year. Meanwhile, certain placental mammals begin to evolve in the direction of man, that is, through the first primate-like forms. Primate history is another chapter in the epic of man's development.

3

The Primate Path to Man

So long as giant reptiles stalked the earth, the earliest mammals quite wisely kept out of sight. With the demise of the great dinosaurs, however, there occurred a sudden and dramatic increase in mammalian forms. A new, milder climate favored the spread of dense forests, both tropical and subtropical, and vast open grasslands. Here the placental mammals would multiply and flourish. Small, generalized animals sufficiently adaptable to withstand the rigors of a changing environment, they fell heir to a land abundant in foodstuffs and bereft of fierce and hungry predators. Many, venturing into the grasslands, gradually increased in size. In time, these would evolve into such ground-dwellers as the elephants, ungulates, and rodents, and into such aquatic mammals as the sea cows and whales.

Others, preferring an arboreal habitat, kept to the trees. Among these were the *insectivores,* primitive insect-eaters who would, in time, found the primate order. For those of us anxious to get on with the story of man's fossil history, it is here that the plot begins to thicken. From these ancient, primitive tree-dwellers are derived all modern primates, from monkey to ape to man himself (Figure 1).

Scientists whose primary interest is the prehistoric development of *Homo sapiens* spend a considerable amount of time studying man's primate relatives. And there are plenty of them to study, for the primate order includes more than 300 different species, living and extinct. Knowledge of their evolutionary development helps us to place man in proper perspective, to understand his nature and trace his origins. Man, after all, *is* man because he belongs to the primate order and thus shared in certain evolutionary trends typical for the primates as a group.

That group is highly diversified, ranging from the primitive, insect-eating tree-shrew to the complex, culture-bearing animal called man, and including a multitude of other animals so different from one another that it's often difficult to find a single characteristic that defines them as a taxonomic group. They are grouped together because they are descended from a mutual ancestor. They shared an evolutionary past. And they have in common numerous anatomical adaptations for living in the trees, compliments of their earliest ancestors. One of these is an enlarged and modified brain characterized by a reduced sense of smell and a pronounced development of the sense of vision.

Most mammals depend heavily upon their ability to sniff out food, vision being of secondary importance. Hence in the typical mammalian brain a large anatomical area is devoted to the sense of smell while a lesser region is reserved for sight. Life in the trees, however, presents quite a different set of problems than does life on the ground. Here the olfactory sense is less important, for juicy smells don't waft about so freely at tree-top level. On the other hand, good vision becomes crucial, both for locating food among clumps of bright foliage and for getting about from branch to branch. Thus most primates—because they got their evolutionary start in an arboreal habitat—exhibit in their brains a reduction in the area devoted to smell and an enlargement of that area devoted to vision.

Another important arboreal adaptation is the opposable thumb. A ground-dwelling animal can get about easily enough without giving much thought to whether he is sure-footed. His weight is dis-

Figure 1. Some of the living members of the order Primate, representing a gradational series which, broadly speaking, links man anatomically with some of the most primitive of placental mammals: (1) Tree shrew; (2) lemur; (3) tarsier; (4) cercopithecoid monkey; (5) chimpanzee; (6) Australian aboriginal. *Courtesy W. E. LeGros Clark and Edinburgh University Press. From* Antecedents of Man, *1959.*

tributed over four feet firmly planted on *terra firma* and—so long as
he does not make a habit of sauntering near the edge of a precipice
—the ground-dweller needn't concern himself with the matter of
movement. An arboreal animal, however, had better worry about
remaining arboreal unless he wants a quick smack on the head as
he hits ground. Because flat feet weren't made for climbing, it's
a safe guess that our earliest primate ancestors weren't flat-footed
for long. It's likely that weight resting upon the foot tended to
force the first digit apart from the others, providing a foot that
curved smartly around a tree-limb. The evolutionary result was
the development of both thumbs and big toes that "opposed" the
remaining digits and yielded the ability to grasp.

If an opposable thumb seems to you like a minor detail, try
this: place your hand flat upon the table beside you. Now lift,
and, without using your thumb, try to pick up a dime or paper-
clip. Try to unbutton the cuff of your shirt, or throw a football,
or pick up a needle and thread it. Next to impossible.

The ability to grasp, made possible by the development of an
opposable thumb and mobile fingers, has been a critical evolutionary step among the higher primates (Figure 2). It helped to
ensure survival in the trees for our arboreal ancestors, and it is a
skill requisite to the manufacture and use of tools which—as we'll
soon see—helped to make man what he is today.

Once the hands are developed in an arboreal animal so that
they are useful for grasping and exploring, the hindlimbs assume
a greater responsibility for supporting the body. The tree-dwelling primate begins to rest on his hindlimbs, which become heavier
and stronger, and to use his hands for touching, holding, feeding,
grooming, and scratching. These are tasks both practical and
pleasurable, and soon the primate begins to spend much of his
time sitting in an upright position. He need no longer sniff at an
interesting object; now he can pick it up and carry it to his nose,
or hold it directly in front of his eyes. The primate sense of
smell—never vastly important in the trees—now becomes even
less so, and the olfactory region of the brain decreases in size still
further. The hands have taken over much of the work of feeding,
so there is no longer a need for large jaws; these grow smaller.
The long snout is also reduced, partly because of the smaller jaw
size but also because of the reduced sense of smell and the enhanced need for visual acuity. With a short snout, the eyes can

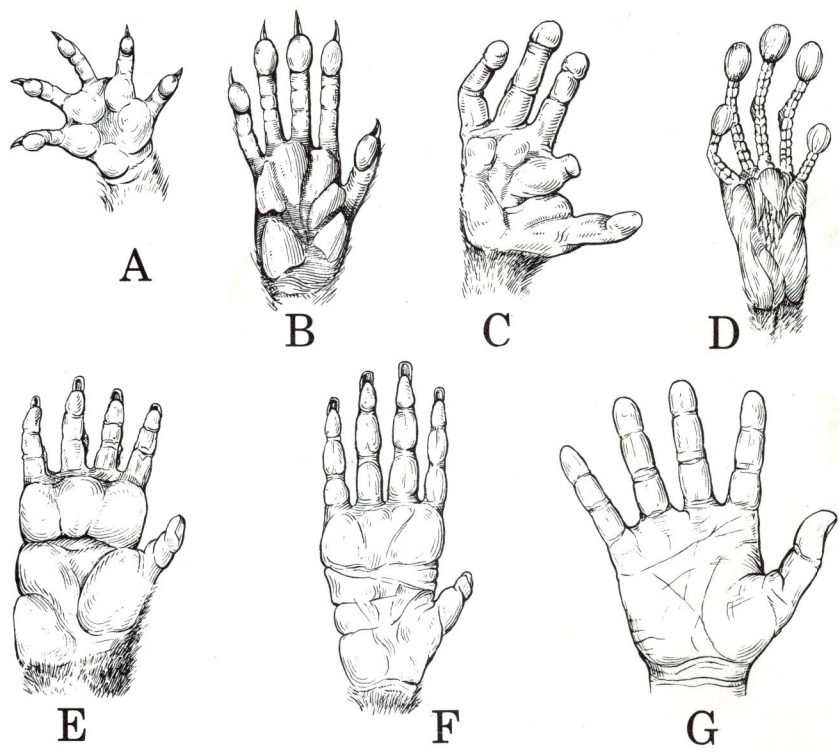

Figure 2. The background of hands:
    A. Opossum—primitive but mobile
    B. Tree shrew—with simple primate proportions
    C. Potto—using thumb and fourth digit for greater span (reduced second digit)
    D. Tarsius—small, with specialized enlarged skin pads
    E. Baboon—used in running as well as handling, fingers somewhat shortened
    F. Orang-utan—lengthened for brachiating, with reduced thumb
    G. Man—with short fingers, long thumb, good opposition

Reproduced by permission of Doubleday & Company, Inc., from the book Mankind in the Making, *by William W. Howells, illustrated by Janis Cirulis. Copyright ©️ 1959, 1967 by William Howells.*

lie close together, improving sight. The primate face begins to flatten out, becoming more "human-like" in appearance, with prominent front-facing eyes, a smallish nose, and a mouth directly under. And these modifications trigger a raft of new adaptations that improve the primate's ability to succeed in the trees.

Not all our modern primates, of course, evolved in an identical manner. No one primate species possesses all the features that tend to arise from similar arboreal adaptations. But all primate species shared, to one extent or another, in the general primate pattern of evolutionary development. And each demonstrates at least a trend toward the evolutionary modifications typical of the primate order.

The primate preference for arboreal living was set in Paleocene times. By then inconspicuous little insectivores had given rise to the first primates to appear in the fossil record. These were the tree-shrews, squirrel-sized creatures with naked tails like those of rats. They haven't evolved much since; modern descendants of the group (Plate 1) look quite similar to their Tertiary ancestors. These are animals so primitive in body structure and brain development that many scientists can't decide whether to group them in with the primates or demote them to insectivore status.

Plate 1. **Tree shrew** (*Tupaia glis belangeri*). *Photo by F. D. Schmidt, courtesy San Diego Zoo.*

Plate 2. Ruffed lemur. *Photo by Ron Garrison, courtesy San Diego Zoo.*

The earliest shrewlike mammals spread and multiplied. From one branch of the family tree evolved the first lemurs, animals hardly more advanced than their shrewish predecessors. Their brains are disappointingly small and almost entirely devoid of the richly convoluted surfaces characteristic for the higher primates. Still, the lemurs have endured for millions of years and are with us today in great diversity. Modern representatives (**Plate 2**) include the shaggy aye-ayes of Madagascar, the sluggish lorises of Asia, and the galagos, often called bush-babies, of Africa. The last are able to curl and uncurl their ears at will, a talent of unknown value (although who can judge except, perhaps, another bush-baby?).

A second main branch of the primitive shrewlike forms evolved in a more exciting direction, that leading ultimately to the tarsiers, monkeys, apes, and man (**Figure 3**). Rare and short-lived in zoos, the modern tarsier is so elusive that few scientists have been able to examine him properly. But he is worthy of both scientific and popular attention: a plump ball of ragged fur with enormous eyes, hairless ears, and a clutch of long, spindly fingers, the tarsier

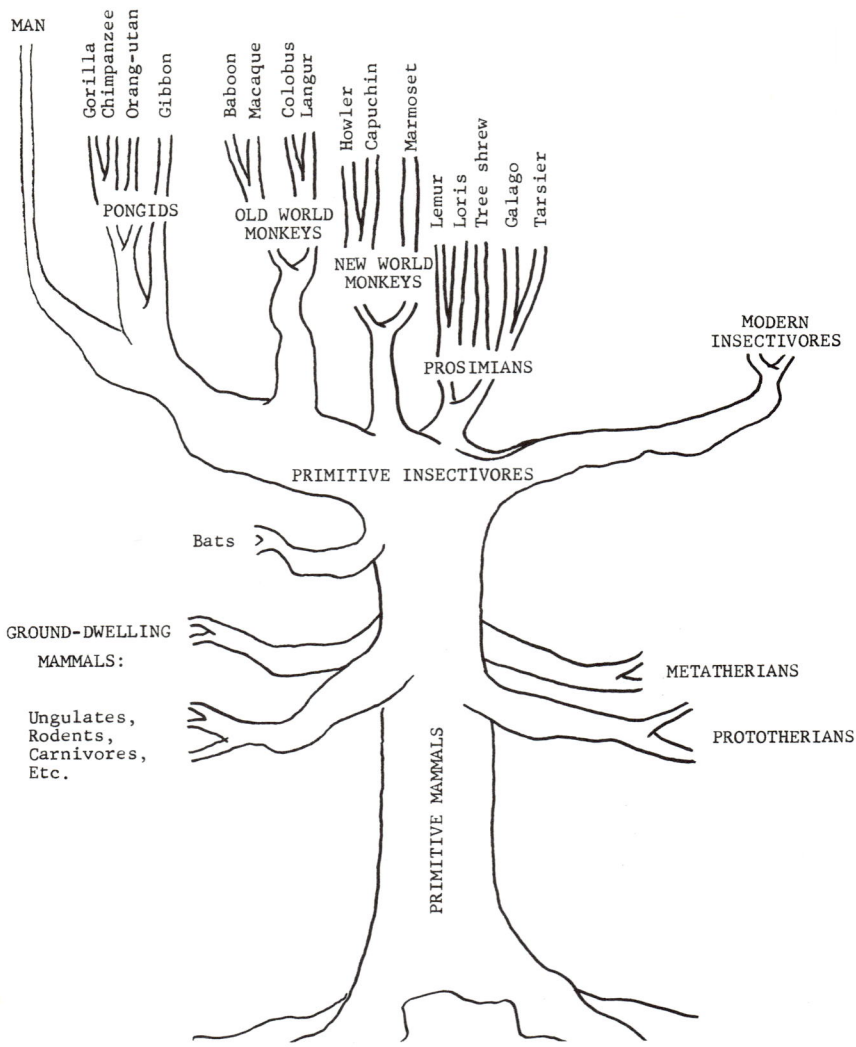

Figure 3. Man's family tree.

(Plate 3) boasts his own curious talents, one of which is celebrated in the following limerick contributed by Dr. Ashley Montagu and sometimes attributed to the late Dr. Earnest Hooton (although neither scientist would acknowledge authorship):

> The Tarsier, weird little beast,
> Can't swivel his eyes in the least.

Plate 3. Tarsier (*Tarsius syrichta carbonarius*). *Photo by Ron Garrison, courtesy San Diego Zoo.*

But when sitting at rest
With his tummy due west
He can screw his head 'round to face east.*

The tarsier has the longest legs, largest ears, and biggest eyes in the primate order, and he's one of the world's first-rate hoppers: he can jump from bough to bough so rapidly that he seems to vanish, only to reappear an instant later on an adjacent limb. But our interest in the tarsier is directed toward his brain. Here we find evidence of an important evolutionary trend: a reduction in the sense of smell and a pronounced development of the sense of vision. By Eocene times, the tarsiers had spread throughout North America and Europe.

* Ashley Montagu, *Man: His First Million Years* (New York: Columbia University Press, 1969), p. 21.

And by the end of the Eocene, the first monkeys had made their debut, quickly separating into two distinct families and reaching their evolutionary peak. One group—the ancestors of our modern New World monkeys—had spread across most of North and South America. A second huge group—ancestral to present-day Old World species—scattered across Africa, Asia, and southern Europe. The differences apparent between these two vast and variable families result from their adaptation to different environments and are preserved by their long geographical separation.

Today's American monkeys, found throughout Central and South America, comprise a diverse collection of forest-dwellers that range in size from that of a kitten to that of a medium-sized collie. Perhaps the most familiar to us is the *cebus* (Plate 4), or "organ-grinder" monkey. But other species—more than 100 of them—abound. All but one are diurnal; they feed in the trees during the day and sleep at night. Highly sociable, they live in territorial groups. They failed to share in the evolutionary devel-

Plate 4. *Cebus,* typical of the New World monkeys. *Photo by S. McKern.*

opment of the opposable thumb. But many species boast a particularly useful arboreal adaptation lacking in the Old World varieties—the prehensile tail. They use this organ almost like an extra hand. Strong enough to hang by, the prehensile tail can be used also as a balancing rudder during movement through the trees. Its sensitive, hairless undersurface is ridged to enhance the gripping function.

A bit more advanced but equally diversified are the Old World monkeys. Their hands, with broad nails and opposable thumbs, are human in shape, and their feet are handlike. Some varieties have cheek-pouches, useful for storing hastily acquired food that cannot immediately be eaten. Others have sacculated stomachs, called "pre-stomachs," which serve the same purpose. Many Old World monkeys are distinguished also by their *ischial callosities,* brightly colored patches of bare, toughened skin developed to accommodate constant sitting or sleeping in the trees.

Most of the Old World monkeys, like those of the New, are arboreal; they feed, sleep, groom, and play in the upper branches. A few—notably the baboon (Plate 5)—have abandoned the trees, however, to pursue a terrestrial existence. To help cope with the dangers of life on the ground, they have developed long, fiercely projecting canine teeth which serve as awesome weapons against predators.

But a more significant baboon adaptation is social rather than anatomical in nature. Baboons live in highly organized troops led by powerful dominant males. The leaders guide troop movements, settle disruptive fights, and—with the help of aggressive co-dominant males—protect the weaker females and juveniles against threats from outside. When the troop is on the move—searching out new food or water sources—individual members arrange themselves in defensive positions. Strong, young males lead the troop and protect the flank. Females and juveniles cluster at the center of the pack. Dominant males remain near the females for maximum protection. Solitary baboons are seldom seen in the wild, for it is only in the troop that an individual baboon can find safety. Modern field studies have yielded surprising new facts regarding the complexity of baboon behavior. Because man, like the baboon, is a terrestrial creature descended from an arboreal ancestor, scientists believe that baboon studies may shed new light

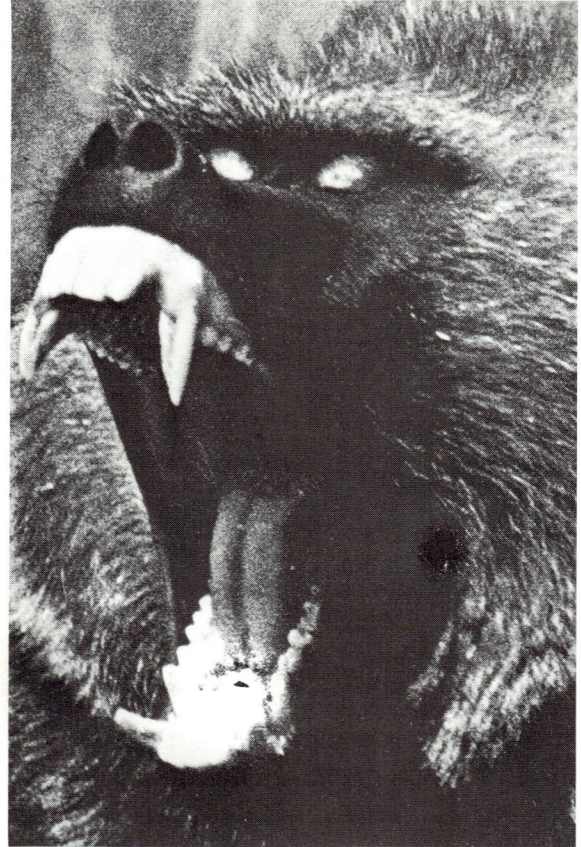

Plate 5. The baboon, an Old World monkey, has adapted to a terrestrial existence. Note the long snout and projecting canines, developed to aid in competition with ground-dwelling animals. *Photo courtesy Claud Bramblett.*

on the development of social behavior among the earliest human groups. But more about this later.

The Eocene was a time of unparalleled primate radiation. Paleocene insectivores had given rise to the first, most primitive primates. Lemurs and tarsiers swept across the globe to dominate the world's forests for millions of years, throughout much of the Eocene. With the appearance of the more advanced monkeys, however, many of these were displaced, their modern descendants limited to Africa and Asia. The monkeys, in dazzling variety, reigned supreme for the remainder of the epoch. Then, toward the very end of the Eocene or at the beginning of the Oligocene, the first apelike forms appeared. These quickly multiplied, spread, and diversified. There was *Parapithecus,* an Egyptian primate no larger than a modern squirrel. *Dryopithecus,* native to Europe, Africa, and Asia, ranged in size from that of a gibbon (about three feet tall) to that of a modern gorilla (five to six feet tall).

And there were giants, too, like *Gigantopithecus,* a creature very much like *Dryopithecus* except that he came in a king-sized package. Our first evidence for the existence of this startling beast came accidentally, when the Dutch paleontologist G. H. R. von Koenigswald, browsing through a Hong Kong drugstore, came upon fossil teeth that seemed to belong to an ape. If so, they represented the most enormous ape ever imagined, for some experts estimate *Gigantopithecus* to have stood twelve feet tall!

For many years controversy raged over the taxonomic status of this immense beast. One American anthropologist, Franz Weidenreich, examined the teeth and decided they belonged to an ancient giant human. Other experts argued that *Gigantopithecus* was no more than a representative of a race of giant apes long extinct. Still others, reluctant to accept the creature as human, nevertheless puzzled over *Gigantopithecus'* seemingly human form of teeth.

More than 1,000 *Gigantopithecus* teeth, together with several jawbones, have since been unearthed. These were recently examined by Dr. David Pilbeam of Yale University, who worked with Doctors Elwyn Simons and Peter Ettel in analyzing these fossils. Dr. Pilbeam suggests that a branch of the *Dryopithecus* group moved millions of years ago from a forest environment into an open-country habitat. A change in diet—to grains and seeds—would explain the more human form of the dentition. Despite the human form of the teeth, however, Dr. Pilbeam classifies *Gigantopithecus* as an ancient ape. There can be no doubt that this beast was the largest ape ever to perch on the primate family tree.

The earliest apelike forms appeared about 30 million years ago, about the time that the monkeys were establishing themselves in forests all across the globe. By Miocene times several different species of true apes inhabited, in abundant numbers, the vast tropical forest that extended from West Africa to the East Indies, an extensive region then uninterrupted by water barriers. These were not brainy or overgrown monkeys, nor were they descended from monkeys. The earliest apes were a group of primates anatomically distinct, their uniqueness arising from a new and different form of locomotion: *brachiation,* that hand-over-hand movement through the trees practiced by pongid tree-dwellers. Monkeys run, on all fours, *atop* tree branches; apes swing by their arms *from* the branches.

Brachiation, made possible by the development of long, mobile fingers, reinforces the habit of upright posture. In brachiating animals, the clavicle or collarbone acts as a strong strut to permit free arm movement in almost any direction; the pelvis is modified to bear the brunt of support in upright stance. All the apes, then, bear evidence of a long history of brachiation: their trunks are short and compact, their arms long and free-swinging. On the ground, they move on all fours but, again, in a manner uniquely pongid. Although they can walk upright like a man, they do so reluctantly and with obvious distaste. Their preferred gait is quadrupedal; they get about by leaning forward to rest their weight on the bent knuckles of their hands.

Numerous attempts have been made to trace the development of the apes from the Miocene to the present. But there occurs from Miocene times one of those annoying voids in the fossil record, a great slice of time for which we have no truly significant fossils. Until we recover additional fossils from that ancient period, a vast chunk of ape history is lost to us. Meanwhile, all we know for certain is that of the numerous prehistoric ape species, only four descendant groups survive today: the gibbons and orang-utans (Plates 6 and 7) of East Asia and the chimpanzees and gorillas (Plates 8 and 9) of Africa. These are man's nearest primate relatives; they share with us numerous similarities in anatomical structure, cellular structure, and blood chemistry. Of them, the gibbon is least manlike; the gorilla—because of his manlike foot—is closest to man.

And what of man? When did he and the apes diverge to make their separate evolutionary ways? The same frustrating gap in the fossil record that obscures ancient ape history also obscures the history of man's earliest development. The earliest manlike fossil

Plate 6. The gibbon, smallest of the living apes, has the longest arms in the pongid family, a result of his long history of brachiation. *Photo by S. McKern.*

Plate 7. Orang-utan (*Pongo pygmaeus*) from Borneo. *Photo by Ron Garrison, courtesy of San Diego Zoo.*

Plate 8. The chimpanzee (*Pan*) is now known to fashion simple tools in the wild.

Plate 9. The gorilla is now judged closest of all the primates to man, primarily on the basis of total blood chemistry and foot morphology. Above, the lowland gorilla (*Gorilla gorilla gorilla*) "Big Man" of the Swope Park Zoo, in Kansas City, Missouri. *Photo by S. McKern.*

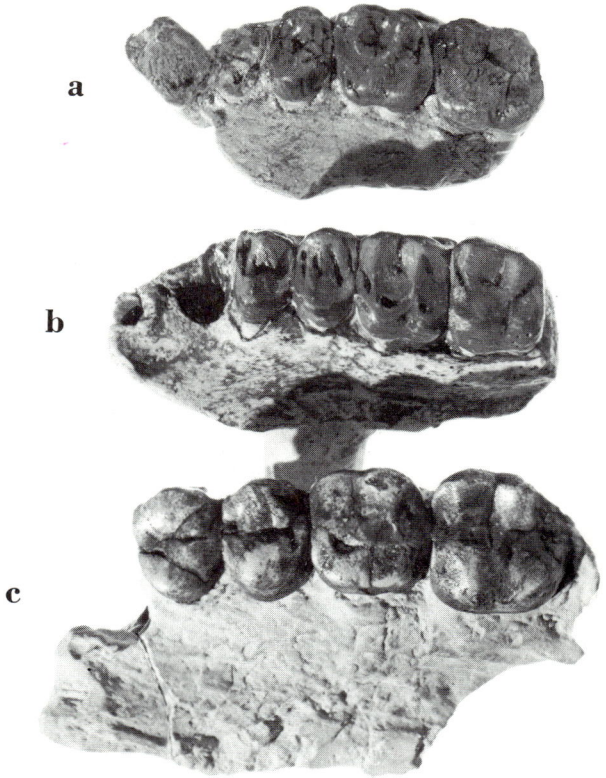

Plate 10. *Ramapithecus,* 8 million to 10 million years old, may represent the earliest hominid form: (a) a fragmentary fossil from a site near Fort Ternan, Kenya; (b) and (c) fragmentary maxillae from Haritalyangar, Simla Hills, India. *Photo courtesy Elwyn Simons.*

so far found is *Ramapithecus* (Plate 10), a fossil 8 million to 12 million years old.

Scientists have long believed that they could identify man's earliest ancestor on the basis of his teeth alone. Apes tend to have long, projecting canines for efficient tearing and shredding of vegetation for food. Man, however, prepares his food by hand or through the use of simple tools—and thus has no need for giant canines. Although this reasoning may fall under attack with the new interpretation of *Gigantopithecus,* it is a sound general rule to follow. And many scientists believe that *Ramapithecus* displays exactly the form of teeth we'd expect to find in the earliest manlike creatures. So far as we know today, then, the primate lineage leading to man diverged at least 8 million years ago—perhaps earlier—from the generalized primate stock that would later produce the modern apes.

One problem in seeking out the earliest human forms arises from the fact that early primates display curious combinations of anatomical traits—characteristics of both the apes and man together with features that seem typical for neither.

When we begin to uncover fossils that seem half-man, half-ape, it's time to stop and decide how to separate the men from the apes. Man is, after all, grouped with the apes in taxonomic classification because he shares with them many anatomical similarities.

But man differs from the apes, too, in numerous important ways. It's easy enough to spot those differences in the living. We're not likely to go about mistaking apes for men or vice versa. But in tracking down man's earliest fossil relatives, we run the risk of capturing fossils difficult to classify. Then we must fall back on skeletal traits which distinguish man from the apes— differences easily obscured when all we have to work with are a few heaps of dusty bones and a handful of crumbling teeth. Our problems multiply when we realize that there was no magic moment in prehistory when man crossed the line from pongid to hominid status. Nevertheless, we must find some way of distinguishing the curious forms we find in ancient deposits.

How do you spot a man in the fossil record? Experts once argued that the only sure way to distinguish ancient men from ancient apes was to find buried with them, in clear and certain association, primitive stone tools. Early anthropologists defined man as a tool-making primate; that is, a form possessing the intelligence and manipulative ability necessary to manufacture tools and weapons. But we mentioned back in the beginning of this book that the search for fossil man centers not only upon excavated evidence but also upon the observation of living forms, particularly of wild monkeys and apes whose behavior helps us to reconstruct the lives of prehistoric men. And just such observations as these have forced a revision of the traditional definition of man.

Jane van Lawick-Goodall, an energetic young primatologist working in the forests of Tanzania, was the first to provide undeniable evidence of what scientists had suspected for quite a while: that wild primates are capable of fashioning crude implements from materials at hand.

Quietly watching her chimpanzee subjects, Mrs. van Lawick-Goodall noticed that a wild chimp would occasionally be attracted to a teeming termite hill. He'd search about for a twig or bit of straw, break it to the proper length, lick it once or twice, and plunge it deep into the termite hill. After waiting a moment—giving the insects a chance to bite into the straw—the chimp would withdraw his simple tool and nibble off the clinging termites.

This is tool manufacture, however primitive, for it demonstrates the chimp's ability to see a problem, decide on a solution, and carry out a procedure that requires the manufacture of a fashioned implement. Laboratory experiments with both chimps and gorillas have confirmed the fact that nonhuman primates can assemble crude tools with or without human direction.

And so anthropologists were forced to devise a new definition of man. Man must now be defined as a primate both intellectually and physically capable of *systematic* tool manufacture through *consistent* methods of production. He makes *patterned* tools and implements. And he is able to teach succeeding generations his techniques of manufacture. The earliest man-made tools known are of stone, with flakes chipped off one or both sides to form a jagged cutting edge—a crude tool but a patterned one, and one not yet duplicated by a nonhuman primate. Some human populations of the distant past are known solely from the distinctive patterned tools they left behind.

We can tell fossil man from fossil ape readily enough—so long as the former was considerate enough to get himself buried with his tool kit. Few prehistoric men were so inclined, and the majority of fossils are found without direct associations with ancient tools. Thus anthropologists and prehistorians must focus upon anatomical or morphological traits that reliably distinguish man from ape in the fossil record.

Nineteenth-century investigators solved the problem by measuring the cranial capacity of each recovered fossil skull. Cranial capacity, of course, refers to the volume of the inside of the cranium; as such, it gives a good estimate of brain size. Comparisons among the primates could thus be made by utilizing the following average values as a rough rule-of-thumb:

	Cubic Centimeters
Modern man	1450
Gorilla	600
Chimpanzee	400
Orang-utan	400
Gibbon	100

Early anthropologists simply dubbed their discoveries human if the brain size were large; if small, they dismissed them as apes.

But a large brain size relative to total body weight is a trait typical not only for man but also for apes. While it's true that humans have a larger cranial capacity than do apes, there is no absolute numerical value that separates man from his primate relatives on the basis of brain size alone. Individuals vary, whether hominid or pongid. And the average cranial capacities noted above are taken from *modern* specimens rather than ancient ones. Nineteenth-century investigators had no way of knowing the average brain size for the earliest humans. In attempting to measure ancient remains by modern yardsticks, they succeeded only in confusing the issue.

More reliable than absolute brain size in distinguishing man from ape in fossilized remains are the relative limb proportions. Because man has a habit of walking erect, his legs tend to be longer than his arms, a state of affairs exactly opposite for the apes, who developed long arms for efficiency in the trees.

Skull features are important, too. In man, the forehead is generally high and the brow-ridges small; in apes, there is no forehead and the brow-ridges are large and bony. Hominid skulls exhibit both chin and nasal bridges; these are generally absent in ape skulls. Man's canine teeth are small; pongid canines are large and projecting, with adjacent gaps in the jaw to accommodate these toothy defense weapons.

Again, because man walks erect, his *foramen magnum*—the opening in the skull where the spinal column meets the brain—is positioned near the center of the skull base. In the ape, whose head tends to hang forward, the *foramen magnum* exists nearer the back of the skull, though not of course so far back as in animals with a more horizontal, or pronograde, posture. Man's dental arcade is parabolic, while that of the ape is more U-shaped (see Figure 4). All these features help to distinguish human from ape remains. And, in addition, modern anthropologists look for *patterns* or *mosaics* of anatomical features, for it is not the presence or absence of a single trait that makes man, but a total configuration of distinctively human morphological characteristics.

 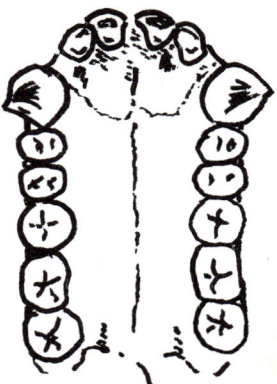

Figure 4. Comparison of dental arcade for upper jaws of man (left) and gorilla (right).

The thing that accounts for these differences is man's unique way of life. He walks erect, his head balanced atop his spinal column. He is omnivorous. He makes and uses tools rather than depending upon such body structures as projecting canine teeth or fierce claws. In hunting, eating, and defending himself, man utilizes cultural rather than anatomical equipment.

The similarities between man and ape have been known for more than a century. They served to support the theory that the two groups shared at some time in the past a mutual primate ancestor. What was not known was the way in which man developed into his present state.

How does an apelike ancestor evolve, through time, to emerge in human form? Until the missing fossil links between man and ape could be located, there could be no convincing justification for Darwin's theory of organic evolution.

But in 1925, a few elusive links began to appear. Intriguing new discoveries in South Africa would hint at the prehistoric existence of strange ape-men (or man-apes)—fossils that would tell of a small-brained primate who dared to abandon the security of life in the forest in order to seek an active and dangerous existence in the grasslands. That, too, is another story.

4

Man-Apes and Ape-Men

Charles Darwin had informed·a skeptical world that man descended from a primate form that lived many thousands of years ago, and that this remote primate was ancestor not only to all modern varieties of man but to all living monkeys and apes as well.

Early misinterpretations of Darwin's theory triggered and sustained the erroneous idea that there somewhere exists a single missing link. Since the initial publication of Darwin's works, a steady stream of exuberant scientists have come forth with their fossil discoveries, each proclaiming his own as the best possible candidate for that unique title.

Darwin himself unknowingly laid the foundation for such a notion when he talked of "connecting links," not yet found, between modern man and his distant primate ancestors. Many reasoned (as some do now) that if man truly evolved from an apelike creature of the darkest past, the scientists could easily prove it by locating the fossilized remains of an archaic form that stands midway between ape and man—hence the popular conception of anthropologists as men engaged in an endless quest for the bones of that single, elusive transitionary form.

This is not to say that no missing links have been discovered. The pages of this book are filled with just such forms. And anthropologists continue to scratch through heaps of ancient soil in the hope that they will uncover still more connecting links.

This is precisely the point: the fossil record is *littered* with the bony remains of transitional forms—not one or two, but a whole series of ancient bones that represent developmental stages through which man passed on his way to becoming human.

Some links, of course, are more critical than others, and the one most anxiously sought since the middle of the nineteenth century was that missing form that would bridge the gap between modern ape and man. It was this urgent desire to find the remains of a creature half-ape, half-man, that led to one of anthropology's greatest and most embarrassing puzzles. In 1912, Charles Dawson, an amateur archaeologist, brought forth fragments of a skull and jaw that seemed to represent the original and long-sought "missing link"—the stage at which man diverged from his pongid relatives. The skull, when reconstructed, was unquestionably human; the jawbone was fully apelike. Named *Eoanthropus dawsoni* ("Dawson's Dawn Man") but popularly called Piltdown man (Plate 11), this discovery rocked the world of science back upon its heels. Some of the world's greatest scholars accepted the find as genuine, jubilantly proclaiming the missing link to be missing no more. Others refused to believe that a jaw so apelike could belong with a skull so human in form. And heated controversy raged for some forty years before modern testing methods proved Piltdown a fraud. By then, Dawson was dead, beyond questioning as to his part in the hoax. But a hoax it clearly was: the skull represents a modern human; the jaw was taken from the ape, its teeth filed down to disguise them.

We have said that Darwin, in supporting his theory of evolution, failed to utilize the fossil material that was rapidly accumulating in his time. Darwin's ideas were firmly based upon his meticulous observation of living forms and of the similarities among them. He was aware, however, that the final proof of his theory lay in the recovery of fossils that showed a progressive

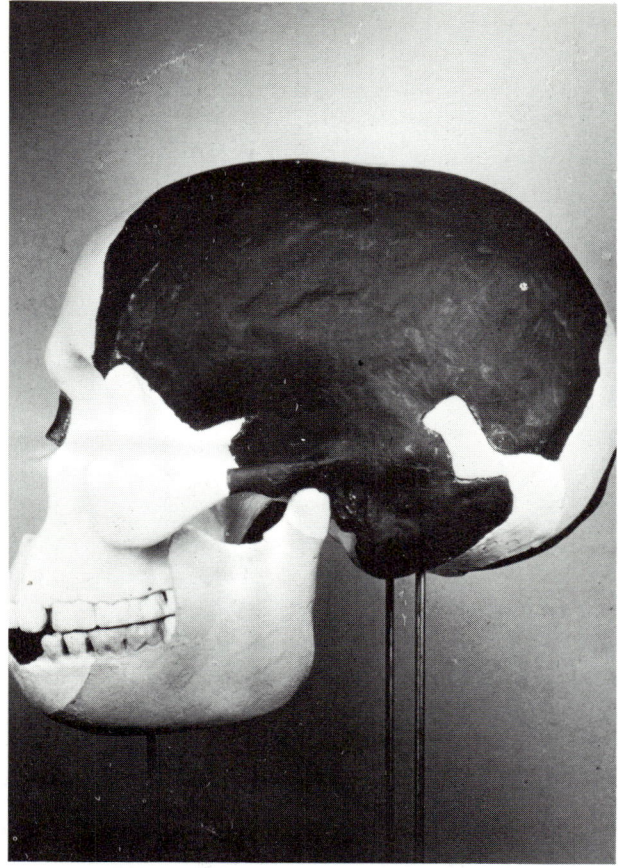

Plate 11. Piltdown man, once believed to represent a crucial missing link, was proved a hoax by modern testing methods. In this reconstruction, black and gray portions represent fragments found; white portions were added by the sculptor to assemble a complete skull. *Photo courtesy Robert Squier.*

development in the direction of man. Darwin predicted with astonishing accuracy that the skeletal evidence for man's most ancient precursor—the one clearly intermediate between modern apes and man—would one day appear somewhere on the continent of Africa.

In prehistoric times, most of the world's land masses were subject to violent upheavals. The earth buckled with the forma-

tion of vast mountain ranges and trembled with the fiery eruption of active volcanos. Four times, catastrophic glacial advances
swept across the globe, smashing forests in their frigid fury. South
Africa, however, escaped these cataclysms. Serene and undisturbed, this land might house the fossilized remains of man's most
distant representatives. And so it was here that many scientists
were drawn in their search for the link that would complete an
evolutionary chain for man. A few spent their entire lives engaged
in that frustrating quest.

It wasn't until 1924—forty-two years after Darwin's death—that
science picked up the trail of early man. Workmen quarrying
lime from deposits near Taung, South Africa, blasted loose from
ancient breccia (cave debris) a small skull that looked strangely
human—but not quite. Puzzled, they mailed it to Raymond A.
Dart, Professor of Anatomy at Johannesburg's Witwatersrand University.

Dart knew at once that this was the fossil so long sought: that
of a man-in-the-making, a creature not yet human but one that
provided proof of the evolutionary leap from anthropoid to
hominid—from ape to pre-man.

The Taung skull (Plate 12) was marvelously complete for so
ancient a fossil. Many parts were available for study. The skeletal
face was almost intact. The lower jaw was present. At least half

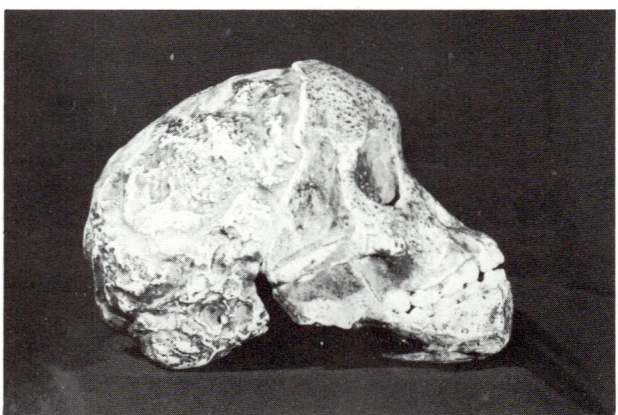

Plate 12. Cast of infant's skull from Taung, South Africa.
Photo courtesy Douglas H. Smith.

the brain cast—an impression of the brain itself upon prehistoric rock—was clearly visible. And there were teeth.

These were deciduous or "milk" teeth: the Taung skull represents a child about six years old at death. And he died in early Pleistocene times—some 800,000 years ago.

The youth of the fossil would spell trouble for Dart. Taxonomic classification is normally based upon *adult* traits. Immature bones merely hint at adult morphology. Further, identifying either an existing animal or a newly discovered fossil on the basis of a single specimen is a risky business at best. Because all living individuals vary, reliable classification is made through analysis of a *group* of specimens—the more the better.

Nevertheless, Dart was satisfied with the conclusions he was able to draw from the single juvenile skull. "The specimen," he wrote, "is of importance because it exhibits an extinct race of apes intermediate between living anthropoids and man." In describing "Taung Baby," Dart made special mention of numerous anatomical features that seemed to him to demonstrate a closer approach to man than they did to any of the living or fossil apes. Since the fossil was decidedly prehuman rather than human, he named it *Australopithecus africanus,* the "South African ape."

Dart had rushed enthusiastically into print, publishing his preliminary conclusions barely six weeks after receiving the skull for analysis. The reaction to his report was immediate and devastating. Both British and American scientists complained that Dart was irresponsible in publishing so radical an opinion without first consulting an older, better-established scientific authority. They ridiculed Dart's attempts to classify an immature specimen and labeled as gross exaggerations his statements regarding the manlike qualities of the skull. "Dart's Baby," the disgruntled experts insisted, was no more than a primitive form of chimpanzee. Even Dart's closest colleagues refused to accept his interpretations. Many wrote disparaging letters. Others published disdainful papers criticizing Dart's rash and hasty examination of the discovery. Dart was enveloped in a cloud of protest, contempt, and ridicule.

The man who came to Dart's rescue was a Scottish medical doctor who had distinguished himself as an extraordinarily competent paleontologist. At an age when most men look forward to a comfortable retirement, Robert Broom—then sixty-eight years old—accepted the demanding position as Curator of Vertebrate Paleontology and Physical Anthropology at the Transvaal Museum at Pretoria, South Africa. After his examination of the Taung skull, Broom was convinced that the world-wide criticism of Dart's work was unfounded. Dart was right, Broom insisted, and the proof that would vindicate him lay in the eventual recovery of additional fossil forms. What was needed to settle the true taxonomic fate of *Australopithecus* was the recovery of mature specimens of the same type. And Broom set out to accomplish just that.

His method was deceptively simple. Broom realized that quarry workers often used dynamite to remove limestone locked in the caves and crevices so common to this section of South Africa. He notified the mining managers of every quarry in the region—a 200-square-mile block of the Transvaal—and instructed each to keep a vigilant eye out for fragments of fossil bone that might be uncovered through blasting. This done, Broom had only to wait and hope.

But he was not a man for whom waiting was an easy task. He made weekly rounds to nearby quarries, naggingly reminding managers of his work and, often, grabbing a shovel to personally attack the mining debris. His assistants might work in khaki shorts and pith helmets, but Broom himself dressed each day in an impeccable, freshly pressed business suit complete with winged collar, stylish four-in-hand, and gold-linked cuffs. Although he worked on his hands and knees scratching through ragged heaps of rubble, he was as unsmudged and unwilted at the end of the day as he was when he set out in the morning.

His remarkable search system began to pay off in August of 1936 when a small fossil brain cast was blasted loose from the breccia at Sterkfontein, about thirty miles from Johannesburg. The resident quarry manager turned it over to Broom, who promptly began to ferret through this new deposit of mining rubble. By noon of the following day he had extracted a skull base, cranial fragments, and large portions of an upper jaw. Once these pieces were assembled, Broom had on his hands a new Australopithecine. This second fossil skull exhibited the same hominid features as its predecessor from Taung, but—most im-

MARKET WEIGHTON COUNTY SECONDARY SCHOOL
EAST RIDING EDUCATION COMMITTEE

portantly—this was an *adult* specimen, one judged by Broom to represent a female.

The Sterkfontein skull was estimated to have a cranial capacity of about 500 cubic centimeters, about the size of the skull of a modern gorilla. The teeth—especially the molars—were large relative to total brain size, clearly a pongid trait. But the general morphology and pattern of these ancient teeth were unmistakably human. The lack of overlapping canines, so characteristic for all apes, indicated that the Sterkfontein individual chewed its food with a lateral movement, producing a surface wear that was flat, as on our teeth. And from the placement of the *foramen magnum,* it was evident that the Sterkfontein man-ape walked in an upright position, his head held erect.

Not only did this skull duplicate many of the characteristics noted in the juvenile skull from Taung, but it also exhibited many of the traits Dart predicted would be present if and when an adult specimen of *Australopithecus* was recovered.

Yet, amazingly, Broom decided that his new fossil discovery was sufficiently different to warrant classification in a separate genus of the same taxonomic family. He christened the Sterkfontein female *Plesianthropus transvaalensis* ("near-man of the Transvaal"), a name that implies a fairly distant relationship to the young *Australopithecus* from Taung.

However the fossils were individually labeled, Dart and Broom now confronted the scientific world with two specimens of a strange new type, one that displayed a fascinating mixture of ape and human traits. These long-hidden fossil links offered mute testimony to the fact that man's presence on earth extended back some incredible 800,000 years into the past. Moreover, they supported Dart's original hypothesis that small, apelike creatures from the Transvaal represented the earliest ancestors of modern mankind. The experts, still cautious but increasingly sympathetic to Dart's case, approached the African fossils with careful objectivity. Were the Australopithecines apelike men, they asked, or manlike apes?

Broom, energetic as always, had no time to waste on such superfluous questions: he was hot on the trail of more fossil evidence. Continuing to visit his caves and quarries on a regular basis, he tipped workmen more generously than he could afford and lugged back to his museum random bits of skull, limbs, and teeth—parts of other Sterkfontein ape-men and of the animals on which they had fed at the dawn of human existence.

Two years following the initial discovery at Sterkfontein, Broom heard that a young schoolboy named Gert Terblanche had collected some curious old bones from a hill near his home. With characteristic vigor, Broom dashed first to the Terblanche farm near Kromdraai (about two miles from the original site at Sterkfontein) and then to the school where Gert was in class. Quickly, he overwhelmed the bewildered principal with insistent pleas that the boy be excused from school. After delivering an impromptu lecture to both students and faculty on the importance of preserving fossil bones, Broom and Gert rushed off to the hill where the boy had found his bony treasures. Within a few days, they had uncovered enough new fossil fragments to enable Broom to reconstruct a third skull.

This one differed in numerous ways from the previous *Australopithecus* skulls. The teeth were larger, the bones heavier and more robust, the jaw more powerful, and the face flatter. In these aspects, the Kromdraai specimen seemed more apelike than the others. In other ways, however, it appeared to be more human, especially in the form of the teeth. Moreover, animal bones found associated with the skull were different from those excavated with the Sterkfontein Australopithecine. Broom decided that the populations from Kromdraai and Sterkfontein had lived at different times. He assigned the Kromdraai fossil to another new genus and species, calling it *Paranthropus robustus* ("robust near-man"). *Paranthropus,* he said, occupied the Kromdraai area at a more recent time than did the gracile *Plesianthropus.*

World War II intervened to halt all archaeological work in the African caves. It wasn't until 1947 that Broom, now eighty years old, resumed his relentless search for the fossilized remains of earliest man. Excavation was accelerated by using dynamite at all sites. This explosive approach to anthropology made up for lost time and yielded Broom numerous fine fossils. It also evoked a predictably explosive reaction from the Historical Monuments Commission, which charged Broom with destroying cave stratigraphy as well as other evidence that might aid in dating the sites.

Undaunted—and with a little help from his influential friends—.

Broom continued his operations. Receiving a permit from the Commission to work at Kromdraai, he was enraged that no mention was made of his work at Sterkfontein. He immediately halted operations at Kromdraai, taking them up at Sterkfontein, where *Plesianthropus* had been found. This illegal excavation produced one of the finest skulls so far recovered, another adult female specimen of *Plesianthropus*.

Encouraged, Broom opened up two new sites, one at Makapansgat and another at Swartkrans. Material literally poured into his laboratory and, as the skeletal material mounted, so did Broom's tendency to assign new fossil names. His final tally went something like this:

Site	Genus	Species
Taung	*Australopithecus*	*africanus*
Sterkfontein	*Plesianthropus*	*transvaalensis*
Kromdraai	*Paranthropus*	*robustus*
Makapansgat	*Australopithecus*	*prometheus*
Swartkrans	*Paranthropus*	*crassidens*
Swartkrans	*Telanthropus*	*capensis*

Such a profusion of names meant that, according to Broom, there had lived within an area of about 200 square miles four different genera and six species of advanced primates, each demonstrating the interesting combination of pongid and hominid traits that would easily qualify them as strong candidates for the position of modern man's most ancient ancestor.

The confusing situation resulting from Broom's indiscriminate taxonomic procedure began to irritate the scientific community. Many experts called for caution at this early stage of discovery. Others, more outspoken, protested that it was absolutely impossible for so many species to live and compete for the same food sources within a single small ecological zone for so long a period of time. Evolution, they insisted, just didn't work that way.

What Broom failed to acknowledge was the inevitability of the occurrence of normal variation. We see today a wide range of variation among the world's populations. We should expect equal or greater variability among man's earliest ancestors. Thus, from the skeletal material Broom collected, modern scientists recognize but a single taxonomic genus (*Australopithecus*), divided into just two species: *Australopithecus africanus* from Taung, Sterkfontein, and Makapansgat, and *Australopithecus robustus* from Kromdraai and Swartkrans.

Robert Broom died on April 6, 1951, at the age of eighty-four. His place in South Africa was taken over by his young assistant, J. T. Robinson. With a determination reminiscent of his predecessor, Robinson continued Broom's work in the Transvaal. Scientists around the world waited with bated breath as Robinson began to add handsomely to the ever-growing stockpile of skeletons that represent the two known *Australopithecus* populations.

But as scientific attention focused upon the efforts of Robinson and the still-active Raymond Dart, material from another part of Africa began to emerge from the rocks.

The new site was Olduvai Gorge in Tanzania, East Africa. The explorer was L. S. B. Leakey (Plate 13), a diligent worker whose name ranks foremost among the great archaeologists and paleontologists of our time. Dr. Leakey, reared in Kenya by English missionary parents, has devoted much of his life to the search for early man in East Africa. Accompanied by his energetic wife, Mary, he has not only uncovered critically important skeletal evidence but constructed careful theoretical interpretations of the remains unearthed by his shovel. Leakey continues to astound and stimulate his contemporary colleagues with his determination and productivity. At the same time—and perhaps more importantly—he takes a personal hand in training the next generation's professional scientists. Despite a staggering workload and a nonstop excavation schedule, Leakey seems never too busy to pause and encourage a visiting anthropologist or to reply, at length, to a querying letter from a student.

Leakey's own story is a study in perseverence. He began operations at Olduvai in the middle of the 1920's, recovering numerous man-made stone tools together with the slaughtered remains of extinct mammals. He succeeded too in uncovering a continuous geologic sequence that included layers representing all of the Pleistocene in East Africa. But for thirty years the fossil bones he sought eluded him.

Plate 13. Dr. L. S. B. Leakey, shown here with animal fossils recovered at Olduvai, Tanzania, has contributed numerous important early-man fossils. *Photograph by Melville Bell Grosvenor © 1966 National Geographic Society.*

Then, on July 17, 1959, workmen recovered a fragmentary but reconstructable skull, one which Leakey immediately recognized as a new Australopithecine specimen. Although the skull clearly resembled the *robustus* species from South Africa, Leakey followed Broom's unfortunate practice of assigning a new name to each fossil discovery: he called the skull *Zinjanthropus boisei.*

The discovery for which Leakey waited so long would prove to be the most remarkable yet. Amazingly, potassium-argon tests showed *Zinjanthropus* to date from 1.75 million years ago—much earlier than anyone imagined the Australopithecines lived. It now began to seem that with every new find, man's past was being pushed further and further back into darkest prehistory.

But *Zinjanthropus* had more to offer than great antiquity. For at the same level as the skull—although *not* in certain association—scientists uncovered very primitive "pebble" tools, rough cutting implements crudely chipped from smooth, water-worn rocks (Figure 5).

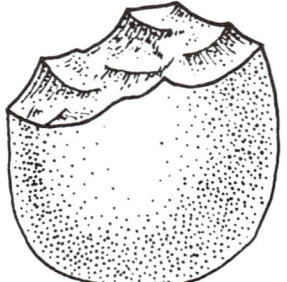 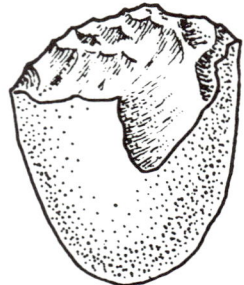

Figure 5. Pebble tools, crudely flaked from water-worn rocks, were the earliest stone tools to be made by man's distant ancestors.

We have seen that tool use and tool manufacture are among the first criteria used by science to recognize the earliest phases of man's development. If *Zinjanthropus* can be proved a true toolmaker, then he must be judged human.

The likelihood that *Zinjanthropus* possessed the intelligence necessary for the manufacture of stone tools remains a matter of heated dispute. All the great apes, of course, are capable of tool use; some, if sufficiently motivated, are known to make implements from available materials. Tool-making, however, requires a certain degree of forethought and imagination, and we would not expect our most ancient ancestors to attempt manufacture unless their way of life demanded it. We can only guess at the events that led to the first deliberate manufacture of stone tools.

Let's count up what we know about *Australopithecus* (Plate 14). We know that he stood erect, his hands freed from the function of support and locomotion. We know that his canine teeth were small and ill-suited for predation and defense. The presence in many of the *Australopithecus* sites of smashed and broken mammal bones is persuasive evidence that he slaughtered animals for food. If so, he must have competed for such food with the fiercer

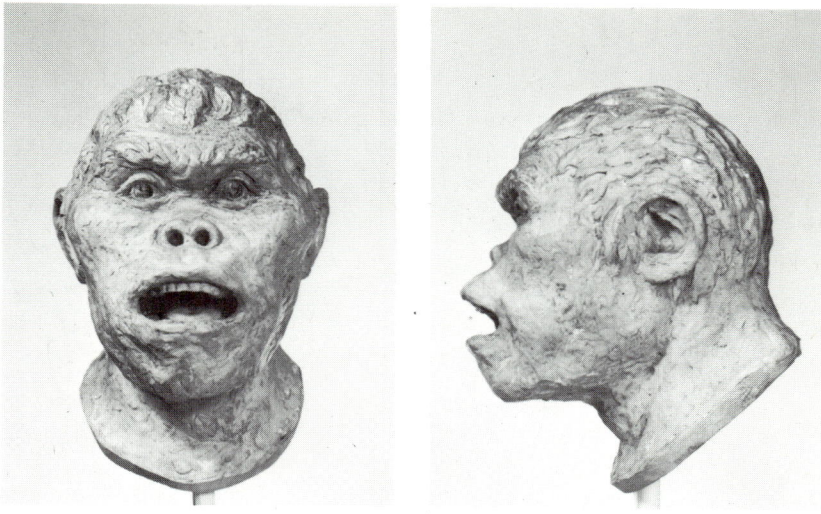

Plate 14. Sculptor's reconstruction of *Australopithecus africanus (formerly Australopithecus prometheus)*. *Photos courtesy of the American Museum of Natural History.*

hunting animals, those endowed by natural selection with swift limbs, strong jaws, projecting fangs, and claws.

Australopithecus must have begun as a vegetarian, living on wild fruits and grasses easily reaped. At some point in his history, however, he acquired a taste for meat. Presumably, he was first a scavenger, robbing the kills accomplished by other, more formidable predators. Perhaps he once hurled a handy rock at an attacking tiger; perhaps—as seems likely from the huge quantities of stacked rock found in East Africa—he was accustomed to defending himself with stones. However he reached the realization, *Australopithecus* must have known at a very early date that he could compete successfully against marauding predators only by manufacturing tools that would compensate for his lack of anatomical ferocity.

The shattered bones found at typical *Australopithecus* sites include those from rabbits, birds, lizards, rodents, and antelope. *Australopithecus* was indeed a meat-eater. And most anthropologists believe now that tool manufacture was within the capability of this ancient man-ape. This conviction has contributed to the hominid status of the Australopithecines and helped to convince many experts that *Australopithecus* played a crucial role in the evolutionary development of man.

Thus Leakey's *Zinjanthropus* find is extraordinary for the light it throws upon a heretofore dark era of human history. It implies not only that tool manufacture began earlier in time than scientists had previously supposed, but also that a creature with a relatively small brain was capable of both forethought and memory—qualities that spell m-a-n for most authorities. We should not be surprised if *Australopithecus* were permitted full hominid status tomorrow.

Spurred on by the reception of *Zinjanthropus* (as well as by increased financial aid), Leakey expanded his African activities. Late in the 1960's, he discovered more Australopithecine remains from an earlier level, represented this time by the fragmentary skeleton of a young adolescent. Despite its rather tender age, Leakey became convinced that these bones demonstrated the existence of a new type of early hominid—one with a slightly larger brain than the neighboring *Australopithecus*. He called it, therefore, *Homo habilis*.

In doing so, Leakey raised a storm of protest not unlike the one touched off earlier by Robert Broom. Critics who accused Broom of recklessly assigning new taxonomic labels on the basis of scant skeletal evidence reacted no more favorably to Leakey's similar tendencies. As a result, the validity of both *Zinjanthropus* and *habilis* fell under attack. Even today experts remain in violent disagreement over the placement of these two fossils in separate and distinct taxons from the now extensive Australopithecine populations.

Leakey, undaunted, remains at work at Olduvai. We can expect further exciting discoveries from this intrepid scientist. But the most recent chapter so far in the history of *Australopithecus* shifts our attention to a small site on an ancient streambed north of Ethiopia's Lake Rudolph. Here, in 1969, an American team headed by Professor F. Clark Howell of the University of Chicago reported the recovery of some forty teeth and two lower jaws, the collection reputedly representing both *Australopithecus africanus* and *Australopithecus robustus*. As yet, excavators have turned up no stone tools in association with the remains. But Omo River deposits, from which the bones were taken, have been assigned a date that approaches 4 million years in antiquity. If this incredible date proves sound after extensive dating methods are applied, the Omo River site could claim a longer evolutionary history for *Australopithecus* than ever has been suspected from the sites in the Transvaal and at Olduvai Gorge.

Assuming—as most experts do—that all of this scattered skeletal material represents a single taxonomic genus, can we reconstruct the life and times of *Australopithecus* as man's earliest known precursor? We can try. Although we may never have a complete picture, we can take the facts gleaned from excavation—the fossil teeth and fragmentary bones—and splice these together with our knowledge of the intelligence and social life of modern primates. The result, like a jigsaw puzzle, grows increasingly more complete as each new discovery adds a missing piece. Some, of course, are lost to us forever. But those we have tell a compelling story.

By the end of the Tertiary—some 3 to 4 million years ago—the former wet-forest conditions had begun to recede. South Africa became extremely arid. Some areas resembled the modern Kalahari desert—barren and dry—while others developed into vast grass savannahs. The disappearance of the great extensive forests, coupled with the change from a moist to a dry climate, forced existing animals to adapt to new living conditions and—most especially —to new and diverse food sources.

Some forms, inextricably tied by ancient patterns to a vanishing way of life, failed to make the change. These were doomed to extinction by slow and agonizing starvation. Others, more versatile in dietary habits, bridged the transition with relative ease. Their survival permitted the passing to succeeding generations of those special features and talents that promote adaptability. Among these fortunate forms were the Australopithecines.

Australopithecus adapted, in fact, not to one new life-way but to many. He thrived as easily in the dry desert conditions of the South as in the arid veldt-lands of the East, both environments dramatically unlike the tropical forest habitats associated with the modern chimpanzees. This capacity of *Australopithecus* to cope with and to exploit diverse living conditions is particularly evident when we consider his total geographic range. So far, probable Australopithecines have been found in such far-flung locales as Ubeidiya, in Jordan, and on the southeastern Asiatic

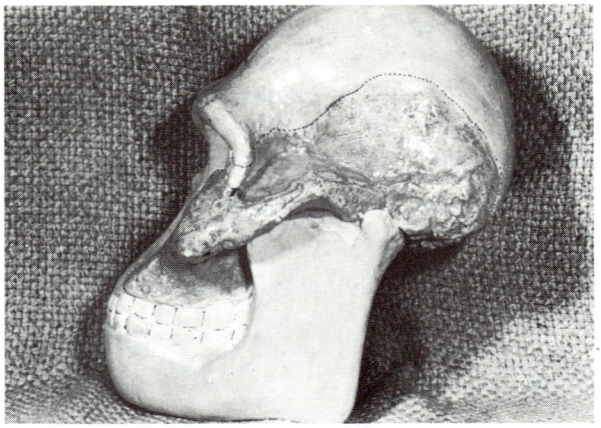

Plate 15. *Australopithecus robustus,* represented here by a cast of *Paranthropus* from Kromdraai, South Africa, was an aberrant form, doomed to extinction. *Photo courtesy University Extension, The University of Wisconsin.*

island of Java. Not only did *Australopithecus* flourish over widespread areas of the Old World, but he spanned too a vast period of time—at least a million years.

What did this man-ape (or ape-man) look like? Specimens collected to date number over 100 individuals, not counting the recent discoveries from Ethiopia. We have, then, sufficient skeletal evidence to sketch a fairly reliable picture.

Compared with modern primates, *Australopithecus robustus* (Plate 15) was a large animal, standing about five feet tall and weighing close to 200 pounds. He had a massive round skull with a cranial capacity of about 600 cubic centimeters.* The skull itself is reminiscent of that of a modern gorilla, with great bony crests and keels for the attachment of heavy temporal and neck muscles. The *robustus* mandible, or lower jaw, was huge and featured strong crushing and grinding teeth. The face was short and wide, with what anthropologists call a medium facial prognathism; the lower face, that is, projected forward. From the structure of the pelvis and the form of lower leg bones, we know that *robustus* stood and walked in an erect position. And from the

---

* It must be remembered that it is the *structure* of the brain, not size alone, that counts when determining differences in brain function. Unfortunately, we have little or no evidence with which to judge the intellectual capacities of these distant fossils.

general morphology and surface wear of his teeth, we infer that he—unlike *africanus*—was essentially a vegetarian, favoring a root-bulbs-fruit diet very similar to that of the modern primates.

Robustus was a large, strong, heavy, slow-moving individual fond of seeking out caves for shelter and protection. Most of his daylight hours were spent hunting food, for in excluding meat from his diet he must have required tremendous amounts of filling vegetation to curb his constant hunger.

Australopithecus africanus (Plate 16) was smaller and more gracile, lacking the heavy muscle markings on the skeleton that typify *robustus*. His skull was long and narrowed. Compared with *robustus,* it was smooth and housed a smaller (500 cubic centimeters) cranial capacity. Like his skull, his face was long and narrow, with some prognathism. His teeth were very much like ours, suggesting that his diet must have been much like that of modern man. Although he can't be termed a gourmet, *africanus* could—and did—eat just about anything. In fact, *africanus* was generally more like a man in all of his characteristics (except size).

The small, agile, intelligent scavenger called *africanus* habitually stood and walked erect; he probably was not capable, however, of the unique striding gait typical for modern man. And he

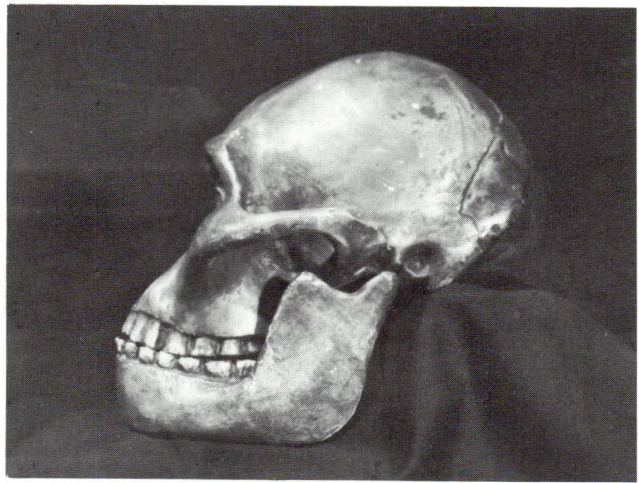

Plate 16. Cast of the skull of *Australopithecus africanus,* the most advanced of the Australopithecines. *Photo courtesy Douglas H. Smith.*

was completely defenseless against the four-footed, ground-dwelling carnivores of his time. He lacked claws, hooves, horns, or projecting canines with which to stand and fight. Nor did he possess the speed necessary to outrun danger. Clearly, he required some other talent to compensate for his many weaknesses.

That talent must have been the ability to fashion with his own hands from stone, wood, or bone crude stone tools and weapons with which he could slay food animals and defend himself against lurking carnivores. From the fossil hand-bones that have been unearthed, there is no doubt that the *africanus* hand was capable of the manipulative movements required in the manufacture of simple tools and implements. Although such tools must have been incredibly crude, they were necessary for the survival of *africanus*. They must have been effectively utilized.

We picture *africanus* living in good-sized groups (again, for protection) in homes of rock shelters or shallow caves; the deeper caves were taken by dangerous predators, and *africanus*, we think, was smart enough to avoid direct and hopeless confrontations if possible. With their simple stone weapons, *africanus* groups hunted and killed small mammals. Young or sick animals of the larger species were probably clubbed or stoned; larger animals may have been driven off cliffs. From time to time, *africanus* scavenged the kills of the fiercer carnivores. Because *africanus* lived in groups and apparently cooperated in hunting activities, he must have developed the beginnings of articulate communication and maintained fairly stable social units.

Obviously, we are speculating beyond our evidence in reconstructing such a life for *Australopithecus*. But chances are good that our speculations range not far afield. We are just beginning to investigate this early history of man's development, and we can look forward to knowing *Australopithecus* better and verifying our present suppositions. Each new discovery adds a piece to our evolutionary puzzle. Soon, perhaps, we can link *Australopithecus* to the next known stage of human development: that of *Homo erectus,* the earliest known undisputed man.

5

Man Emerges

Some fossil discoveries are made by chance, others by design. The first fossil men to come to light in modern times appeared—as we will see in Chapter 6—as an unexpected by-product of Europe's booming nineteenth-century industrialization. Widespread construction of new roads, buildings, mines, and quarries ravaged the surface of the land and uncovered, quite at random, the first puzzling traces of early man. These were the bones of Neanderthal.

On the other hand, it was no accident that Robert Broom encountered *Australopithecus* in South Africa. Broom knew that this was the region most likely to yield up the bones of man's earliest ancestors. Dart's fragmentary man-ape skeleton had narrowed the field, pinpointing the location Broom must search. Broom's quest, then, was deliberate. He stalked *Australopithecus* in serene confidence, never doubting that sooner or later he would find the bony evidence he sought.

Broom's conviction that Africa presented a prime excavation

site was hardly unique. Forty years before the appearance of the first *Australopithecus* skull, a young Dutchman named Eugene Dubois dreamed of finding in Africa primitive fossils that would help to illuminate the darkest gaps in man's prehistory.

Dubois was born in 1858, his earliest years coinciding with the initial period of fossil discovery in Europe. In those days the infant science of anthropology was taking its first wobbly steps; furor and controversy greeted the announcement of each new fossil find. Dubois was determined to be a part of this great excitement. He had prepared at first for a career in medicine. But, increasingly, his interests shifted to the fields of anatomy and natural history, the sciences that form the foundation for the study of human evolution. By the age of nineteen, Dubois had become obsessed with the idea of finding a truly primitive and apelike fossil that would validate Darwin's radical new theory of evolution as it applied to man. Later, as Lecturer in Anatomy at the University of Amsterdam, he was constantly distracted from his work by visions of the sensational discoveries awaiting him in Africa—if only he could get there! It was on the African continent that man's closest animal relatives—the chimpanzees and gorillas—survived in great numbers, and in the African forests that the ancestors of these advanced primates must have arisen. Surely, Dubois reasoned, man originated there also. And if this were true, meticulous excavation must certainly result in the dramatic recovery of fossils previously unimagined. It was inconceivable to him that someone else should have the satisfaction of finding them.

But for Dubois—young, inexperienced, lacking wealth, reputation, or specialized training in paleontology—an African expedition lay beyond reach. At first reluctantly and then with renewed hope, he turned to the sites he considered next best. The Asiatic islands of Java and Sumatra, like South Africa, had escaped the rigors of prehistoric glacial advance. Here, too, advanced primates dwelled: dwindling orang-utan populations persisted in the dense forests of Java and Borneo, a neighboring island once connected to the Asiatic mainland. An abundance of extinct animal bones had been found on Java, suggesting that game had been plentiful even during times of bitter cold. There was a chance—a good one—that early man had inhabited these islands also. If so, his fossilized bones must lurk there still, hidden in the earth's secret depths. Dubois had only to search them out.

To the shocked dismay of his friends, Dubois readily tossed aside his promising career as a teacher. Resigning his post at the University, he wangled an appointment as Surgeon in the Royal Dutch Army. His reason? This position would take him to Sumatra, where he would at last have a crack at locating the bones of earliest man.

Once settled, Dubois attacked his hospital duties with vigor, dispatching them as rapidly as possible so that he might devote every spare moment to his quest. Every hour he could steal away from the operating room was spent roaming the countryside, digging in fossil-rich stream-beds, scouring the island's many caves.

Frustrating weeks passed with nothing to show for his endless grubbing but a handful of teeth from an extinct variety of orangutan. Then came word from a neighboring island that human bones—old ones—had been found. Dubois shamelessly bombarded the Army with urgent appeals. It was essential, he insisted, that the Dutch government have an accurate record of extinct vertebrate fauna native to its island territories; his training in anatomy and natural history would enable him to submit a superior report. When his request for transfer was granted (perhaps as a means of halting Dubois' endless flow of arguments), Dubois rushed to Wadjak, on the southern coast of Java.

He wasted little time there. He purchased the skull found by others earlier and, almost immediately, unearthed another. Both seemed disappointingly recent: he was on the trail of the most ancient man and would settle for no less. He packed the skulls away and kept digging. Working near Trinil, a small village on the banks of Java's Solo River, he recovered a fragment of jawbone—prehuman, he thought—with a single molar tooth still in place. The deposits in which he was digging were rich in fossils. Some 350 feet deep, their stratigraphy dated back to Tertiary times. Dubois was certain that he would find, at any moment, the bones of a crucial missing link.

To his despair, the rainy season intervened to make further excavation impossible. Impatiently, Dubois paced with his fragmentary jawbone and waited for the rains to pass. On the first clear day he resumed his muddy search. Soon he unearthed an apelike molar, and then a handful more. A month later, he stumbled upon the prize for which he had risked his career: a single heavily fossilized fragmentary skull.

Mostly braincase, the skull was unlike any seen before. Too large and heavy to represent the skull of an ape, it was nevertheless too small to be that of a man. The forehead was low and primitive; huge bony brow-ridges projected in a bar across the front, lending a brutish apelike appearance. It seemed to Dubois a perfect transition form, one that bridged the gap between ape and man. Excitedly, he plotted large-scale excavations.

But maddeningly, the rains resumed, halting operations once more. Again Dubois waited, resenting each wasted second. Then, digging at the same level as before—but about forty-five feet distant—he recovered a complete femur, or thigh-bone. Amazingly, it appeared fully human in form, hardly distinguishable from that of modern man. After agonizing appraisal, Dubois decided that it belonged with the previously excavated skull.

Dubois had gone looking for the earliest man. Now, it seemed, he had found something even better: a beast half-man, half-ape. Announcing his discovery of a "human-like transition form from Java," he named the fossil *Pithecanthropus erectus* ("erect ape-man"), the name chosen by an earlier scientist to designate the hypothetical creature linking man and ape—if and when such a form could be found. *Pithecanthropus,* said Dubois, stood erect, some 5 feet 8 inches tall as estimated from the length of the femur. His cranial capacity (between 900 and 1,000 cubic centimeters) was well beyond the range known for modern apes but within the lowest allowable limits for man. His teeth were a curious mixture of pongid and hominid traits, a state of affairs to be expected in a transition form. And he lived in late Pliocene times—an incredibly remote date for a prehuman.

The scientific community received Dubois' report with open skepticism. Dubois, after all, was merely an amateur paleontologist making a preposterous claim for a mismatched heap of fossils found on an out-of-the-way island. And if science was hostile, the Church was aghast. *Pithecanthropus,* complained an irate and unified clergy, made a singularly unattractive Adam. And as a missing link, *Pithecanthropus* was unacceptable on religious grounds.

Dubois, unable to uncover any further evidence along the Solo River, hastened to Europe to defend his discovery. Exhibiting the controversial bones at scientific meetings at Paris, London, Berlin, Edinburgh, and Dublin, he fielded angry questions from assembled experts. How could he be sure, they demanded to know, that the skull and femur belonged to a single creature? How could he account for so modern a thigh-bone in so ancient a geologic deposit? A few disgruntled scientists waged vigorous campaigns to convince Dubois that *Pithecanthropus* was no more than a large chimpanzee. Others, scoffing, dismissed it as abnormal, a pathological freak of nature accidentally preserved from the distant past.

Humiliated, outraged, stinging from the verbal assault, Dubois removed his precious fossils from public view. Burying them in a strongbox beneath his dining-room floor, he vowed he'd have no more to do with science or scientists: he was done with the lot of them!

It soon became obvious that Dubois had blundered in assigning a Pliocene date to his fossil ape-man. Dubois' work in Java had aroused the interest of the German zoologist Emil Selenka, who organized, in 1906, a second expedition to this Asiatic island. Selenka died suddenly the night before departure, leaving his work in the capable hands of his widow. Madame Selenka failed to uncover further traces of *Pithecanthropus,* but she added to the value of the fossil by examining more scientifically the geologic deposits from which *Pithecanthropus* had emerged. On the basis of extinct animal remains, she recognized Djetis, Trinil, and Ngandong levels as representing early, middle, and late Pleistocene deposits. The antiquity of *Pithecanthropus,* taken from Trinil levels, was immediately revised from Pliocene to mid-Pleistocene times, a date more acceptable to those who opposed Dubois' report. Dubois, still in virtual hiding, did not comment.

Then came a discovery that would tempt Dubois to break his long silence. In 1918, a skull excavated earlier at Talgai (near Queensland, Australia) came to the attention of trained paleontologists. A highly fossilized skull representing an adolescent male, the Talgai fossil looked remarkably like a hypothetical ancestor to the present-day Australian aborigines.

Dubois, not to be outdone, threw open his strongboxes and

brought forth the skull he had excavated at Wadjak—before his discovery of *Pithecanthropus*—and the skull he had purchased on his arrival on Java. He had failed to exhibit these skulls in the past, initially because he was on the track of older, more primitive game and later because he wished no "lesser" discoveries to detract from his proud presentation of *Pithecanthropus*. But now, he boasted, he would show that both Wadjak fossils demonstrated a striking resemblance to the skulls of living Australians. Probable ancestral forms, it seemed, did have a way of turning up after all.

Dubois' Wadjak skulls, together with the fossil from Talgai, provoked great excitement in the scientific community. And they led to a renewed interest in *Pithecanthropus*. If the Wadjak skulls were valid, wondered the world's leading scientists, why not have another look at the *Pithecanthropus* bones?

But Dubois, still smarting from the rude reception he'd received with his Solo River ape-man, obstinately refused to expose himself (or his treasured fossil) to further humiliation. He rejected urgent appeals from the presidents of both the American Museum of Natural History and the prestigeous Dutch Academy of Sciences. But finally he relented, permitting Ales Hrdlicka, Curator of Physical Anthropology at America's Smithsonian Institution, to examine the long-hidden bones. Hrdlicka even managed to persuade the irascible Dubois to show the remains once more before a scientific audience. Viewing the fossils a second time, the experts were more responsive, deciding that *Pithecanthropus* represented an "erect ape-man" after all, a transition form advanced far beyond its pongid ancestors.

But by now important new fossils had begun to emerge from China. These would provide evidence to vindicate both the peevish Dubois and his apelike friend. It is an ancient tradition in China that "dragon bones" (which are, in fact, the fossilized remains of many different sorts of animals) can be ground up and made into powerful potions that cure almost any ailment from heart-burn to lovesickness. Since 1903, when the German scientist Max Schlosser purchased a Pleistocene molar tooth in a Peking chemist's shop, paleontologists had made it a habit to browse in Chinese apothecary shops.

The Chinese government, distrustful of all foreigners, permitted scant excavation by outsiders. But Western scientists, intrigued by the "dragon bones" that kept turning up in Chinese drugstores, suspected that, were they able to excavate the hills surrounding Peking, they might very well encounter valuable skeletal evidence concerning early man. This lingering suspicion grew into certainty when a Swedish geologist named J. G. Andersson, acting as Mining Advisor to the Chinese government, came across the first tantalizing clue to Chinese prehistory—a clue that would lead to the discovery of one of the richest and most productive archaeological sites ever known.

Conducting an extensive geologic survey of the limestone cliffs near Choukoutien—about thirty miles south of Peking—Andersson noticed among other rocks broken bits of quartz, a material not native to the region. Surely these had been brought in by human hands.

But what man would transport such quantities of raw quartz—and for what purposes?

For Andersson, there seemed but one compelling answer: blocks of quartz had been carried to Choukoutien in the remote past by primitive stone-age men who fashioned weapons and tools from stone, leaving behind these discarded flakes. Andersson ordered immediate excavations in the hope of locating the bones of these lost tool-makers.

He was soon joined by Davidson Black, a Canadian anatomist working at Peking Union Medical College. Black was convinced that mankind had originated in Asia. He had roamed both Inner Mongolia and Siam for the bones of prehistoric man. Now, agreeing with Andersson, he would sift all of Choukoutien, if necessary, to find the fossils that would prove man arose in Asia. But his investigation, he decided, must be thorough. He made plaster casts of all the Asian fossils so far uncovered (including *Pithecanthropus*) and carried them to the Rockefeller Foundation, where he pleaded for sufficient funds to conduct an exhaustive study. His high-keyed enthusiasm for the project was contagious: not only did he receive a grant from the Foundation, but he convinced even China's reticent officials that he should proceed—with the understanding, of course, that the Chinese government would retain possession of any valuable specimens recovered.

Assured of financial support and governmental cooperation, Black broke ground in April of 1927. Within six months his field director had unearthed a massive molar tooth. It was both huge and primitive, but undeniably human in form—so human, in fact, that Black did not hesitate to use it to create a special new hominid genus. He called this new form of man—represented by but a single molar—*Sinanthropus pekinensis* ("China Man from Peking").

Two years passed before the first appearance of a *Sinanthropus* skull, found by the eminent Chinese paleontologist W. C. Pei. But this was followed by a veritable flood of fragmentary fossils. Exuberantly the scientists pressed on, exposing at Locality 1 a dramatic scene of prehistoric brutality. Here rested thirty-eight broken skeletons: their skulls were shattered, their long bones split apart.

Peking man, it seems, was a cannibal. He had ravaged the bones of his kin for brain and marrow.

The workers at Choukoutien felt no revulsion, readily overlooking this character flaw. For, to the unbounded joy of those digging in the hills surrounding Peking, *Sinanthropus* had left behind a wealth of material through which he could be known. There were crude stone tools and fire-blackened rock—clear evidence that *Sinanthropus* was a tool-maker sufficiently advanced to harness fire for his own uses, more than half a million years ago.

The skull of *Sinanthropus* is so strikingly similar to that of *Pithecanthropus* that it settled the taxonomic fate of the Javanese specimen. If *Sinanthropus* possessed the intelligence necessary for tool manufacture, so must have its near-twin *Pithecanthropus,* for the differences in skull and braincase were minute. No doubt remained now that Dubois' "erect ape-man" was in fact the earliest known true human. Dr. Black joyously flashed the word to Dubois. But the unpredictable Dutch eccentric suddenly reversed his original interpretation. *Pithecanthropus,* he insisted now, was neither ape-man nor full-fledged human but an extinct variety of giant gibbon; the newly discovered *Sinanthropus* was no more than a degenerated Neanderthal inexplicably strayed to China. There was no relationship between the two fossils.

Pithecanthropus, clearly, would have to take his rightful place in prehistory without the support of his erratic discoverer.

Back in Java, *Pithecanthropus* had a new champion. The young German paleontologist G. H. R. von Koenigswald had been dispatched to seek out and recover additional specimens of this ancient hominid.

Von Koenigswald began his excavations along the Solo River, hoping to find any fossils inadvertently overlooked by Dubois. Soon he had collected eleven strangely smashed skulls, broken remains that would later be identified as the victims of another prehistoric cannibalistic feast. These would prove to be significant fossils, but they were not sufficiently ancient to represent *Pithecanthropus,* the hard-sought target of von Koenigswald's hunt.

He pushed on. At Modjokerto, on Java's east coast, he exposed a crumbling infant's skull. At Sangiran, he recovered an adult specimen of the same type, a form like *Pithecanthropus* but a little more primitive in skull features. These he named *Pithecanthropus modjokertensis,* assigning them to early Pleistocene times.

In 1937, von Koenigswald's appointment came to an end and his job was abolished. But he would not abandon his search. He rushed to the United States, persuaded the Carnegie Institute to finance further excavations at Java, and set to work anew. He returned to Sangiran, a collapsed slope west of Trinil, to prove it the most important early man site in Java. Facing its vast expanse of stony rubble, however, von Koenigswald despaired at the thought of total excavation: a thorough search of this immense slope might prove an impossible undertaking.

Then he hit upon what he thought was a brilliant strategy: rounding up a group of unemployed natives, he promised 10 cents for each fossil man fragment recovered from the rocky rubble. He sat back, grinning broadly at this sudden stroke of genius. But his period of self-congratulation was short-lived. Soon he was confronted by a happy mob of excited workers, each grasping a tiny bit of bone. His fossil corps had found a human skull. But, thinking to fatten their pay envelopes, had broken its larger fragments into smaller pieces.

Von Koenigswald counted this a lesson well-learned and set about the tedious task of reconstructing the shattered fragments. When he was done, he held in his hands a new and nearly complete *Pithecanthropus* skull. Elated, he rushed the news of his find, including a photograph of the assembled skull, to his friend Dubois. Dubois, predictably unpredictable by now on the subject of *Pithecanthropus,* promptly published the photograph together with a scathing statement disavowing the discovery. The skull, he clearly implied, was a fake. And so ended von Koenigswald's friendship with the man he hoped to emulate.

Meantime, back in China, workers at Choukoutien continued to accumulate *Sinanthropus* bones. Dr. Black, encouraged by the worldwide acceptance of *Sinanthropus,* had resumed excavations with renewed vigor, insisting upon doing everything himself. Black loved his job and seemed genuinely amazed at his good fortune in being paid for doing the work he most enjoyed. But he set too stringent a task for himself: not only did he direct all field operations, but he labored day and night in his laboratory drafting detailed reports of his progress, measuring and describing each new fossil fragment, arranging for multitudinous drawings and photographs, editing constant press releases, plotting future excavations. He had little time for eating or sleeping and little patience for those who wasted time on such pursuits. If he left his office at all, it was to show a visitor proudly about his site. Black suffered a mild heart attack in 1933; he considered this no more than a bothersome minor ailment, ignoring it as nonchalantly as he disregarded the painful effects of the silicosis he had contracted as a result of continuous exposure to drilling dust. But in 1934, weak with exhaustion, Black surrendered to an overburdened heart. He died suddenly in his laboratory surrounded by his beloved fossils.

Black's successor at Choukoutien, selected the following year, was Franz Weidenreich. Born in Germany, Weidenreich had completed his medical training only to find himself irresistibly drawn to the study of human evolution. Politically outspoken, he lost his post at Alsace-Lorrain when the French assumed possession of that territory in the final days of World War I. Later, he was appointed Professor of Anthropology at the University of Frankfurt, only to be driven from that position by Hitler, whose notions on human race ran counter to Weidenreich's teachings of racial equality. Weidenreich was serving as Professor of Anatomy at the University of Chicago when the Rockefeller Foundation beckoned him to China.

Delighted to be at the center of anthropological action, Weidenreich became a whirlwind of activity. Immediately he made exact casts of all the Asian fossil material, a detail that would prove invaluable to later scientists for, during the darkest days of the second World War, the original material would be lost. He initiated a comprehensive analysis of all the fossil bones, assembling the most detailed reports ever completed of Peking man. And then he turned his attention to new researches at Choukoutien.

Workers had begun systematic excavations of the upper levels of Choukoutien in 1933, hoping to find bones that would link *Sinanthropus* with later residents of Asia. In the recesses of an upper cave, they stumbled suddenly upon the grisly scene of a prehistoric mass murder. Seven skeletons—probably the members of a single family—lay in dusty disarrangement, their few weapons and beads scattered about them. One skull—that of an adult male —had been pierced by an arrow. The others were brutally crushed, apparently by blows from a heavy, blunt instrument. There was no evidence of cannibalism. But the tenants of this long-sealed cave most certainly had been the victims of sudden violence at the hands of unknown intruders.

These are not representatives of *Sinanthropus;* the upper cave dates from 20,000 to 30,000 years ago, long after Peking man disappeared from China. But who were these hapless people?

Weidenreich examined each tattered skull in turn, finally identifying among them *three* distinct racial types! The adult male, he said, demonstrated European traits; he may have been Ainu, an ancient race of Caucasoid whose remnants live today in Japan. One female skull exhibited skeletal characteristics typical among modern Eskimos. And a second adult female looked decidedly Melanesian. How could Weidenreich account for the presence of an Eskimo, a Melanesian, and a European together in a prehistoric Chinese cave? He decided that mankind had differentiated into distinct races early in prehistory, perhaps about the time of *Sinanthropus*. The murder victims of Upper Choukoutien reflected a chance mixture of races.

Like so many others, Weidenreich had overlooked the facts of human variability. Modern peoples differ in all physical features; it should be expected that prehistoric peoples varied similarly. Scientists today recognize the human remains from the upper cave as early Mongoloids, their individual differences due to normal human variation. Because the skull and teeth of several of the skeletons are reminiscent of *Sinanthropus,* many anthropologists believe that Peking man gave rise to the modern Mongoloid races.

Numerous early anthropologists stubbed their toes on this matter of human variation. Only a tiny percentage of living creatures become fossilized or otherwise preserved, and we are not often fortunate enough to recover large numbers of a single type. As a result, our taxonomic classifications are based upon only a few specimens—numbers seldom sufficient to distinguish between traits common to a species as a whole and those resulting from individual variation. Imagine for a moment the difficulties that would arise should future archaeologists attempt to learn the physical traits of our modern populations from a handful of random skeletons!

Living forms vary not only as individuals but also as members of local populations isolated from other groups of the same species. The gorilla once ranged Africa as a single continuous population; now the species *Gorilla gorilla* is divided into two "races": *gorilla,* the lowland variety, and *beringei,* the mountain type. Subspecies variations appear whenever a group evolves in isolation from others of its own kind. And it is the recognition of this fact that has led in recent years to a reappraisal of the Java and Peking fossils.

When fossils are discovered, investigators are most often impressed with their extreme features, those that distinguish them from other known fossils. *Sinanthropus,* so similar to *Pithecanthropus,* was initially assigned to a separate genus due to its slightly more advanced skeleton and its larger brain. Later, experts realized that Peking man and Java man possessed more similarities than they exhibited differences. And today they are separated only at the subspecies level.

Most authorities now lump together all the human fossil material from Java and China within a single species, *Homo erectus* (Plate 17), that form generally conceded to represent the earliest known human.* Regional *erectus* populations differed, of course, in certain physical traits and in their life-ways, but it is assumed that all were sufficiently similar to permit interbreeding, which is

---

* At least until, as we anticipate, the Australopithecines are reclassified, on the basis of increasingly widespread opinion that *Australopithecus* was a tool-maker. It is likely that *Australopithecus* will be rechristened *Homo australopithecus,* a step that would make *Homo erectus* the second-oldest true hominid so far recovered.

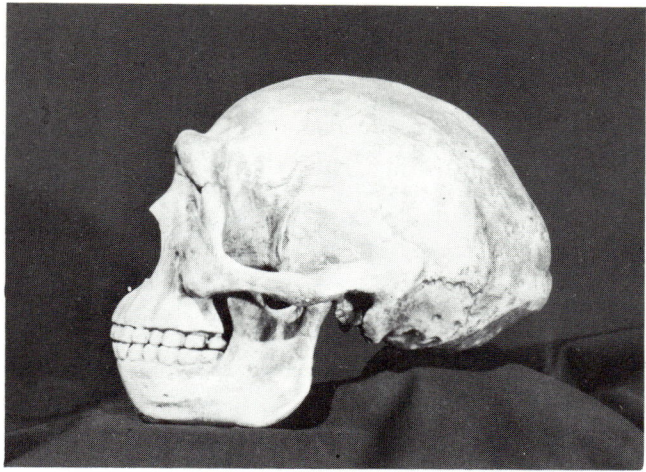

Plate 17. The reconstructed skull of *Homo erectus,* found in far-ranging regions of the Old World. *Photo courtesy Douglas H. Smith.*

the test to determine whether a population can be included within a given species. Anthropologists recognize for *erectus* a number of local varieties:

Homo erectus erectus
Homo erectus modjokertensis (also known as *Homo erectus robustus*)
Homo erectus pekinensis
Homo erectus lantianensis
Homo erectus mauritanicus
Homo erectus leakeyi

Homo erectus erectus is the name given to those fossils previously called *Pithecanthropus erectus,* the ones found by Dubois and, later, von Koenigswald in Trinil deposits. What do we know about Java man?

We know that evolution had favored his post-cranial skeleton, bringing his limbs almost up to modern standards while permitting his skull to lag behind. Java man walked erect half a million years ago and stood some 5 feet 6 inches to 5 feet 8 inches tall—about the height of our average modern man. His limb-bones are barely distinguishable from ours. From the neck down, then, *erectus erectus* boasted a modern form. And the placement on his skull of the *foramen magnum* indicates that his head poised upright upon his spinal column.

But his skull was thick and heavy; his profile, brutish by present-day standards. His forehead was so low that it was all but obscured. Bony ridges loomed heavy above his eyes, lending an apelike appearance, and he lacked the refined, pointed chin so typical for modern humans.

We have no way of deducing the quality of his brain. Several anatomists profess to note a reasonable expansion of the frontal lobe, that portion of the brain associated with the power of speech. And so we assume that Java man spoke, although we doubt that he engaged in prolonged philosophical discussions.

Java man represents a large step—but only a step—away from his pongid ancestors, a kinship reflected in the pongid features of his face and jaws. He lacks the simian shelf (a bony ledge in the interior of the typical ape jaw) but at least one *erectus erectus* specimen displays a diastema, or space, between upper incisor and canine teeth—a trait found in apes to accommodate a long and projecting lower canine.

Homo erectus erectus (Plate 18) was miserly in the clues he left behind him. Just as we can't reliably deduce the mentality of Java man, so are we unable to reconstruct fully his way of life. We do know that he fed upon extinct varieties of elephant, hippo, deer, and antelope, wisely choosing the weak or young as the chief target of his primitive hunt. He probably gathered nuts, roots, and berries—when available—to supplement his protein diet. We can guess that he lived in small family bands, groups seldom exceeding thirty individuals. For the available game in a single region could not have supported a larger population of early hunters. If Java man possessed tools (and we believe that he did, although none have been found in certain association with his bones), these were crudely chipped from stone; he had not yet learned to fashion more sophisticated implements. And he had little time for such niceties: his was not a comfortable or a long life. He lived in endless combat against the elements, predators, and disease. It is estimated on the basis of skeletal material re-

Plate 18. *Homo erectus erectus,* or "Java man," his life reconstructed here in a diorama from the Illinois State Museum. *Photo courtesy Charles W. Hodge and the Illinois State Museum.*

covered that 80 per cent of *erectus* individuals failed to live past the age of forty.

The infant skull unearthed by von Koenigswald at Modjokerto and the adult specimen taken from Sangiran now represent the subspecies *Homo erectus modjokertensis,* a population about which we know even less. *Modjokertensis,* similar to *erectus* but more primitive in all skeletal features, predated Java man in Asia. Many experts believe that *modjokertensis* roamed Java during the early Pleistocene; others, less certain of the authenticity of the geologic dating methods used, place him nearer Java man in time.

For *Homo erectus pekinensis,* formerly *Sinanthropus,* we have a healthier stockpile of information. Peking man (Plate 19) looked very much like his Javanese cousin except that he seems more advanced. His skull was larger (1100 to 1200 cubic centimeters), he had a slight bridge to his nose and the merest hint of a chin, and his bony brow-ridges—still enormous—projected below a defi-

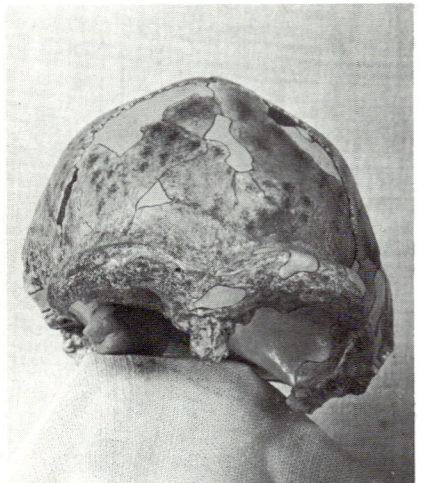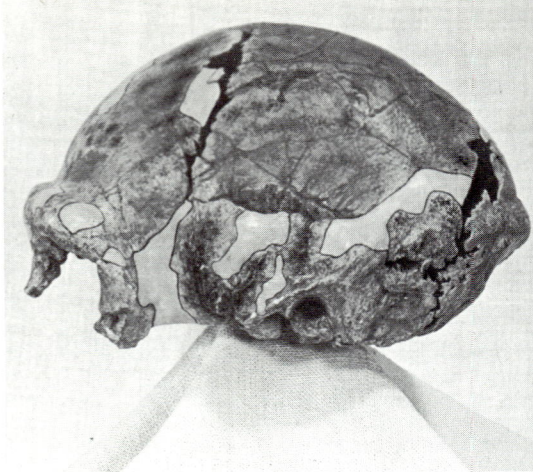

Plate 19. Cast of skull of *Homo erectus pekinensis*—"Peking man"—left, front view; right, lateral view. *Photo courtesy D. Gentry Steele.*

nite trace of a forehead. His jaw lacked the primitive diastema noted among some Java man specimens. One characteristic of Peking man was entirely new: he had *genial tubercles,* the small projections of bone on the front inner jaw to which the tongue muscles are attached in modern man. For many experts, these constitute tangible proof that Peking man possessed the power of speech.

Certainly *pekinensis* enjoyed some cultural advantages unattained by his more brutish cousins. He had learned the benefits to be derived from community living (Plate 20) and had mastered fire, that wondrous discovery that yielded warmth, protection from predators, and a means for cooking meat. His tools were primitive, but from chunks of chert, quartz, and sandstone he managed to manufacture a wide range of crude choppers and scrapers. Judging from these implements, Peking man was right-handed. He was also a cannibal, though some experts—reluctant to make such a charge against one of their own species—maintain the possibility that Peking man was not cannibalistic at all but the victim of someone who was. We can never know for certain.

Nor can we re-examine the original *Sinanthropus* bones for additional information. These were irretrievably lost during the

Plate 20. A cultural reconstruction of *Homo erectus pekinensis,* formerly known as *Sinanthropus pekinensis,* or "Peking man." *Photo courtesy Charles W. Hodge and the Illinois State Museum.*

chaotic days of December, 1941. Dr. Wong, Director of the Chinese Geological Survey, became concerned for the irreplaceable fossils when he heard rumors of a Japanese invasion. Ordering the bones dispatched to the United States for safekeeping until the end of the war, he arranged their transport by railroad to a secret rendezvous with the liner U.S.S. *President Harrison,* the ship that would carry them to safety in America.

Something happened on the way to the boat. The fossil-laden train arrived at Chinwangtao on December 7, the day of the surprise attack on Pearl Harbor. Confusion reigned supreme. The *President Harrison* was deliberately grounded in order to avoid capture by the Japanese; the train was boarded and ransacked by Japanese troops. The bones were never seen again.

Some believe that the fossils had been transferred to the ship, were inadvertently dumped, and now rest lost and encrusted on the floor of the ocean. Others maintain that the Japanese, failing to recognize the importance of the bones, tossed them off the train, after which they were stolen by local Chinese who ground them up and sold them as curative "dragon bones." Still others insist that the fossils were confiscated by the Japanese and remain hidden by that conquered government. Extensive searches at the close of World War II failed to turn up the bones, although equipment carried by the ransacked train was eventually recovered. The Chinese government, angered to this day by the loss of bones legally theirs, maintains now that the original fossils were stolen by the government of Japan, re-stolen by the Americans, and carried stealthily to the American Museum of Natural History—where they remain today, a constant frustration to their rightful owners. This is sheer propaganda: we do not have the bones of *Sinanthropus.* The sad truth is that *Homo erectus pekinensis,* recovered from the earth after half a million years' hiding, has lost his way once more. All that remains of him are Weidenreich's excellent skull casts and a single original molar tooth.

Does this mean that we can never learn more about China's earliest man? Not at all—for new material has begun to emerge from that distant land. In 1963, Chinese scientists uncovered a fossil lower jaw at Chenchiawo, Lantian District, Shensi Province. In the following year there followed face and skull fragments from nearby Kungwangling Hill. The cranial capacity of the reconstructed bits is estimated at 780 cubic centimeters, and the bones are thick, heavy, and more primitive than either Peking or Java man. Native to China before the rise of Peking man, Lantian man—*Homo erectus lantianensis*—may well represent the grandfather to *Sinanthropus.* We must wait and hope for more information from China.

It is apparent now that by mid-Pleistocene times a major human stock had arisen to roam all of eastern Asia and beyond, for we find the fossilized remains of *Homo erectus* not only in Java and China but also in North Africa. In 1955, three jaws and a skull bone were found at Ternifine, Algeria, beneath sand and clay at the floor of an ancient pond. The bones are remarkably similar to *pekinensis.* First called *Atlanthropus mauritanicus,* they are known today as *Homo erectus mauritanicus,* a new piece in the jigsaw that is *erectus*—that prehistoric form which managed to range free across vast reaches of the Old World.

In 1960, the ever-watchful L. S. B. Leakey found a new fossil at Olduvai Gorge—this time a skull-cap dating from 500,000 years ago, at the time that belongs to *Homo erectus.* Excavated from layers above the remains of *Homo habilis,* the skull-cap first was called "Chellean Man," an informal designation based upon its clear association with stone tools of the Chellean type. Its features, however, seem reminiscent of *erectus.* No one knows just what role this ancient fossil plays in the saga of man. But experts dub it *Homo erectus leakeyi* in honor of that energetic discoverer.

From Hungary, in 1965, came a single skull fragment dating within the time period allowed for *erectus.* A particularly puzzling fossil, it is one of the few human bones dating from mid-Pleistocene times that has turned up in Europe. The discovery is so unusual, in fact, that anthropologists can't settle on a suitable name; they refer to the fragment simply as the "Vértesszöllös fossil." Is it *Homo erectus?* No one knows—yet.

Where did *erectus* come from? Where did he go? How did he make his way, on foot, to such far points of the globe? These are questions for which we have no answers. Some scientists, noting that the earliest representatives of *erectus* appeared before the last of the Australopithecines gave way to extinction, venture the guess that it was *Australopithecus* that gave rise to *Homo erectus.* Others argue that the first forms from Java appeared too soon upon the heels of *Australopithecus* to have been so derived. Perhaps tomorrow science will reap a new harvest—the bones of a yet-unknown form predating both these fossil populations, or an intermediate link between them. Fossils have a way of popping up before the ink dries upon the pages of a book summarizing them.

For now, we make do with partial answers, awaiting further skeletal discoveries that will give us fuller ones. The search for early man is a maze of half-facts and glimmering probabilities—a maze richly pocketed with blind alleys.

As in a hall of mirrors, we move slowly.

6

The Neanderthals

For all the excitement *Homo erectus* engenders as our first undisputed true man, he poses almost as many problems as he solves. The clues are there, it's true: thanks to the well-defined stratigraphy of Java's Solo River and China's caves at Choukoutien, scientists are able to assign reliable dates for the *erectus* skeletons excavated so far.

Beyond this, we are permitted only the most tantalizing glimpses of *erectus* life. The discovery of fire-blackened rock in association with his bony remains tells us that *Homo erectus* used fire. We can imagine him hunched over a heap of blazing logs, snatching from the flames crisp hunks of charred meat—and toting with him wherever he traveled a handful of glowing coals encased in a rough stone container, so that he would not have to labor over the start of a new fire.

From the charred animal bones unearthed, we can deduce the typical *erectus* diet. And because he generally slaughtered ice-age mammals for food, we know that this first man from Java lived in a time of bitter cold, although he enjoyed seasonal periods of warmer weather just as we do today. He left behind a nice stock-pile of crude stone tools, often fashioned from material alien to his immediate homeland. We can imagine his seasonal trek from Choukoutien to the hills beyond to gather the workable quartz needed for new weapons. And this in itself is a clue that *erectus* possessed the power of speech, for verbal communication must certainly have been necessary in training the young to produce the crude but surely patterned stone tools of that distant era.

But many frustrating gaps remain. We can't yet be sure where *Homo erectus* came from, or how he evolved. We don't know what forces directed the rapid evolution of his limbs while allow-ing his skull and facial skeleton to lag behind as remnants of his pongid past. We can't date his discovery of fire, or tell how it came about: was he witness to a terrifying bolt of lightning that set a prehistoric tree ablaze—or did a spark from his stone-on-stone approach to tool manufacture ignite a scrap of tinder? We're simply unable to deduce much about his life habits beyond the simple assumption that he lived as a seasonal nomad, search-ing for wild game and gathering what wild fruits and edible berries he found along the way. We don't yet know what became of *erectus*. Perhaps we never shall, for much of his history is mired in the mystery of time forever passed. We will never know him well. We know only that he lived for a time and then disappeared, to be replaced some hundred thousand years later by a more modern type of man—a type we know much better.

No one had to go looking for Neanderthal. He came unbidden, his bony remains turning up all over Europe from the middle of the nineteenth century—long before European scientists were pre-pared to find them. By 1840, European industry was booming. New roads were needed to hasten transportation of manufactured goods; new buildings were erected, extensive mines and quarries opened. In the path of this great new construction, the first traces of fossil man began to appear.

The first Neanderthal skull was found in 1848 in a quarry on the northern face of the Rock of Gibraltar. With it were recovered the bones of extinct mammals and some roughly chipped stone, which should have hinted at a prehistoric origin. The skull itself was a strange one, remarkable for its receding forehead, enormous eye orbits, and gigantic face. But it stirred little excitement among British scientists assembled in London for their annual meetings.

The Gibraltar skull was found too soon.

Australopithecus still slept in the secret depths of Africa's rocky Transvaal.

The man who would stalk Java in search of *erectus* had not yet been born.

Charles Darwin—a quiet, unknown young man weakened by a blood disease contracted during his long and fruitful voyage on the *Beagle*—was still pondering the zoological wonders he had observed in South America and on the Galapagos, still sifting and sorting the pieces of nature's grandest puzzle. He would not venture into print to stun the world with his revolutionary ideas about the development of living forms for another decade.

English scientists on the whole were not interested in the questions of man's origins. Indeed, for them there were no questions that wanted answering. They had been taught and they believed that all the world's living organisms—including man—had been separately and divinely created. Species were immutable, fixed in their present forms from the moment of God's creation. If anyone —scientist or clergyman—wondered why the Creator should see fit to place upon the earth life in such infinite variety, he felt no compulsion to ponder the riddle aloud. The act of creation clearly lay outside the realm of reason and beyond the reach of scientific verification.

And so it was that the Gibraltar skull—the first representative ever to be unearthed of a prehistoric population—was regarded as a mere curiosity to be packed away in a museum where it lay forgotten for more than sixty years.

But Neanderthal was determined to make himself known. In 1856, quarrymen working the limestone cliffs above the Neander Valley in western Germany prepared a cave for blasting. As they dug, they exposed a heap of bones. Failing to recognize them as human, they tossed them aside.

Luckily, the owner of the quarry retrieved the jumbled skeleton and brought it to the attention of Johann Carl Fuhlrott, a science teacher interested in paleontology. By the time Fuhlrott acquired the bones, many of the smaller ones had been lost, together with all the teeth. But the remaining portions—arm and leg bones, rib fragments, shoulder and pelvis pieces, and a complete skull-cap minus mandible—were enough to convince Fuhlrott that there was something extraordinary about these bones.

Quickly, he summoned Hermann Schaaffhausen, an eminent anatomy professor at the University of Bonn. Together they pored over the bones, quizzically examining the thick skull vault and massive brow-ridges. At last Schaaffhausen decided that the Neander Valley skeleton could not belong to any known race of modern man: it must represent a primitive form from the past.

Schaaffhausen and Fuhlrott bundled up their prize and carried it to various scientific meetings, pointing to this "most bestial of all human skulls" and arguing for a validation of their view that the skeleton was an ancient one.

But the time was not yet ripe.

Scholars could not yet comprehend the extent of man's vast antiquity. They had no notion of the true age of the earth, no understanding of extinction. Refusing to accept the concept of evolutionary change, they were unwilling to believe that the Neander Valley specimen predated modern man.

And neither Fuhlrott nor Schaaffhausen could offer proof to the contrary. Because of the casual manner in which the skeleton was recovered—unceremoniously dumped, as it was, from a limestone cave—it was impossible to reconstruct the stratigraphy or determine geologic age. Worse, the skeleton itself was grotesque: the skull was low-domed and thick, the thigh bones curved and heavy. It seemed to resemble an ape more than a man. And few people were willing to admit kinship with this uncomely, ill-proportioned beast.

Better to call it deformed—and numerous scientists did just that, judging the bony remains to represent some unfortunate, diseased form of modern man. One prominent French scholar, his national prejudices visible, estimated the skull to belong to a modern Irishman with "low mental organization." Rudolf Virchow, the leading German anthropologist of his time, pronounced it, quite simply, a pathological idiot.

And this second Neanderthal was dismissed as unworthy of notice by the most eminent scholars of Europe.

But a revolution was in making. Crude stone tools appeared again and again in deposits dating from antediluvian times—and often in unmistakable association with extinct animal bones. More importantly, the science of geology was rapidly coming of age, particularly in France.

Here early geologists noted what we know today: that sedimentary rocks, through the force of gravity, form beds laid down on land or in water, and that these strata are arranged in temporal sequence. Geologists soon formulated the *Law of Superposition,* the concept that in undisturbed layers time is indicated by the sequence of deposit. For the first time, they held the key to geologic time—the key that makes possible our growing knowledge of fossil forms.

French geologists, viewing the graded changes that seemed to be taking place in animal forms with the passage of time, were forced to admit that change in time did indeed occur, a fact that sharply contradicted the common belief that the world and all its life forms were immutable. But this admission served at first only to enhance the theory of special creation.

There had to be some way of reconciling the new knowledge of geology with the doctrine of divine creation. How indeed were the geologists to explain the fossils taken from undisturbed layers —those fossils that indicated an increasing complexity of form through time? It was simple: Baron Georges Cuvier, most prominent exponent of a doctrine known as catastrophism, hypothesized that the fossils found in ancient earth layers *proved* that in the past the world was inhabited by animals unlike those we know today. He did not believe, as we do now, that present forms represent modifications of older forms. On the contrary, Cuvier reasoned that the earth had been subjected to a series of great catastrophies—at least twenty-seven of them—each of which devastated all existing populations. As the earth recovered from each cataclysm, there followed a new divine creation by which new types of plants and animals arose to repopulate the earth. Extinct animal forms, obviously, dated from previous creations, while living organisms dated from the last and were preserved through the final catastrophe, the Biblical deluge. Man himself, according to the catastrophists, had no prehistory because he did not exist prior to the final Creation: that occurred at precisely nine o'clock on the morning of October 23 in the year 4004 B.C. And, because man had no prehistory, he could not be expected to leave his bones in prehistoric earth layers.

Cuvier and his followers enjoyed enormous popularity in Europe; the cataclysm doctrine served nicely to explain the strange, annoying fossils that kept reappearing to embarrass those who held to the special Creation theory of life's origins. But unknowingly the catastrophists were paving the way toward an ultimate acceptance of evolutionary theory. Dedicated as they were to the scientific principles of geology, they were destined to provide the most definitive proof of human evolution.

In 1887—thirty years after the first Neanderthal find—there came a discovery that would vindicate Schaaffhausen and deal a mortal blow to the theory of catastrophism. At Spy Cave in Belgium, two nearly complete fossil skeletons were exposed. Their skulls duplicated almost exactly that from the Neander Valley cave.

Virchow clung to his original diagnosis: such grotesque and ugly forms could belong only to pathological idiots. But, this time, conditions were perfect for authenticating the evidence. The cave stratigraphy was undisturbed, and the geologic age of the skeletons —about 45,000 years before the present—was unquestionable. It was straining coincidence too far to suggest the presence of yet two more fossilized lunatics; the experts—excluding the obstinate Virchow—agreed at last that these skeletons could belong only to members of a prehistoric human population.

For the first time, science admitted the existence of prehistoric men.

Now there could begin organized explorations for the bony remains of early man, and expeditions were chartered in a frenzy of excitement. Scientists scattered, shovels in hand, hot on the trail of Neanderthal man. Obligingly, Neanderthal appeared.

At La Naulette, scientific explorations ordered by the Belgian Government revealed the first Neanderthal mandible, or lower jaw. Found in a pit with the fossilized bones of mammoth, rhino, and reindeer, the jaw differed from that of modern man in the startling fact that it had no chin. For scientists bent on finding a "missing link," the jaw was a happy find. Said Paul Broca, founder of the first anthropological society: "I have no hesitation in saying that the jaw from La Naulette is the first evidence to provide the Darwinists with an anatomical argument."

In 1908, from a cave in France known as La Chapelle aux Saints, came the first complete Neanderthal skeleton—and the best preserved thus far. With the full skeleton, excavators found the remains of such cold-weather fauna as bison, hyena, and reindeer. These were mixed with crude stone tools—points and scrapers of worked flint.

The skeleton, laid on its side with knees flexed, represented an adult male about forty-five years old at the time of his death— elderly by Neanderthal standards. His skull was huge: the brain capacity is estimated to be more than 1,600 cubic centimeters, far beyond the average range for modern man. This Neanderthal, like the others, was short—about 5 feet 4 inches—and boasted a squat, robust body build.

He was no beauty. His limbs were bowed and heavy. Obviously, he walked with a slouch—knees bent, head hung forward, feet plodding. He had known agonizing pain: all his molars had been lost before death, and the teeth remaining were marked by advanced decay and severe abcesses. Still, he must have seemed a lovely sight to those who had tracked him to this lonely cave.

The Old Man of La Chapelle appeared to have been deliberately buried, suggesting that it was Neanderthal who invented religion. Certainly his is the first population known to have ceremonially treated its dead. Exciting new possibilities loomed for those who sought to know this ancient man. Burial implies an awareness of death as well as a belief in an afterlife. Did this hulking, gorilla-like beast know religion?

Further excavations would settle the question. At Le Moustier, five feet below the cave floor, workmen found the remains of a sixteen-year-old boy, obviously prominent in his tribe—for he was buried with an extraordinary number of finely worked tools. His crushed skull lay on a carefully arranged mound of small stones. At his side were the charred bones of a prehistoric ox—food, perhaps, for the journey into eternity. From other sites came additional proof of Neanderthal's observance of rite and ritual: rings of cave-bear skulls surround burial after burial in Europe and the Mid-East.

Now the discoveries came more rapidly than they could be recorded. Multiple burials emerged again and again. From La Ferrassie, in southern France, excavators took an adult male Neanderthal, then a female. In 1912, four children's skeletons followed from the same site. At La Quina, *twenty* complete

Neanderthal skeletons were recovered, a veritable treasure for scientists who had so long been content with teasing bits and tatters of ancient bone. Now the Gibraltar skull was brought out of its museum wrappings and re-examined: it differed not at all from the newer Neanderthal discoveries.

The pieces were beginning to fall into place now. Exuberant scientists armed with ever-growing samples began to sketch a portrait of Neanderthal man. Neanderthal, it seemed, lived in Europe some 45,000 years ago, at the peak of the Wurm I glacial advance. In this archaic world of ice and snow, he survived by joining together into small hunting bands and stalking the huge cold-weather mammals of that distant era: the mammoth, woolly rhino, bison, hyena.

Pitted against a hostile, frigid environment and such fierce predators as the giant cave-bear and sabre-toothed tiger, Neanderthal lived by his wits. He developed an advanced stone tool industry (called *Mousterian;* see Figure 6). With these and a shrewd native intelligence, he slaughtered the largest of the available game, taking their hides to fashion into clothing, thus fending off the awesome cold.

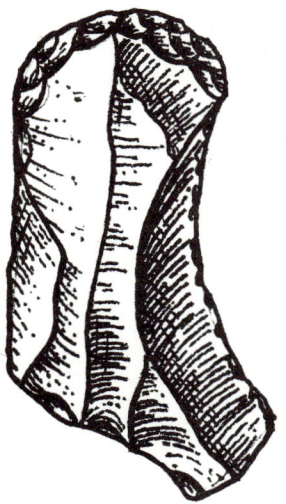 

Figure 6. Typical Mousterian tools: (a) end-scraper, (b) point. The Mousterian tool kit, used by Neanderthal man, includes a high number of scrapers, heavy-duty implements useful for cleaning hides.

From the caves he drove the bears, taking their dens as his own. Inside the limestone caves and rock-shelters, he built hearths by digging out the stone floors; he mastered the sure control of fire, using it not only for warmth and protection but also for cooking. Neanderthal man kept to the front of the cave, reserving its inner reaches for ceremonial purposes. In the depths of his cave he stored skulls and bones taken from the cave-bear, using them perhaps in the making of hunting magic. Between hunts, he huddled beside the fire near the mouth of the cave, his hands seldom idle. From pieces of flint, he chipped small heart-shaped hand-axes, triangular points, spears, and side-scrapers. From animal bones, he fashioned the simple penetrating tools we call *bone awls*. His handiwork is so distinctive in style that scientists can recognize Neanderthal sites from the tools alone, when skeletal remains are not preserved.

Physically, Neanderthal man (Plate 21) resembled nothing so much as a standing bear. He was short and barrel-chested, with powerful forearms and sturdy, bowed legs. His head was huge and hung heavily forward, with a bony projection at the rear which made the skull seem both long and low. Neanderthal had an immense face dominated by a low forehead and receding chin; the massive bony ridges across his brow must have given the impres-

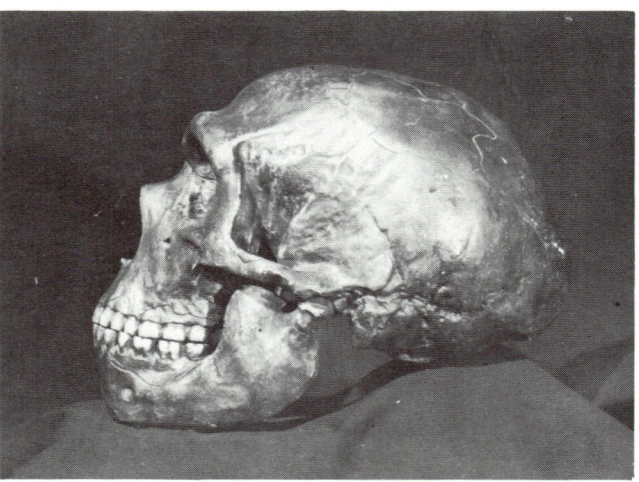

Plate 21. Cast of *Homo sapiens neanderthalensis,* typical of the population in Europe. *Photo courtesy Douglas H. Smith.*

sion of a perpetual scowl. And he had good reason to frown: he was plagued by bad teeth and aching arthritis. Early anthropologists pictured him as primitive, hairy, and brutish—the sort one would prefer not to meet in a darkened alley.

For all his rugged appearance, however, Neanderthal (Plate 22) seems to have been a rather pious fellow. He is the first known man to bury his dead. Now, it is true that, at a few sites, he simply tossed the deceased members of his family into the refuse heap along with spoiled flint tools and gnawed animal bones. But at others, he treated the dead body with red ochre made from earth minerals, laying it out carefully and burying it in laboriously excavated trenches. Into the grave with the body went worked flint tools, animal horns and hides, and curious small objects that could only have served supernatural purposes. Often, he included ceremonial offerings of food; sometimes, he cast wildflowers into the grave before filling it.

Plate 22. Cultural reconstruction of the life of *Homo sapiens neanderthalensis,* or "Neanderthal man," from a diorama at the Museum of Natural History, the University of Kansas. *Photo courtesy Larry G. Quade.*

By 1924, European scientists had begun to feel at home with Neanderthal. Each discovery seemed to fit a new piece into an ever-clearer jigsaw. And each new find was comforting in that it served to authenticate earlier discoveries. Neanderthal man seemed the same wherever he appeared. The geologic dates of all sites were compatible. The skeletal remains constituted a homogeneous population. The cultural associations were identical.

But as the search for more Neanderthals overran the boundaries of Europe, nagging doubts arose. In Russia, at Kiik-Koba in the Crimea, two skeletons—an adult and a child—were recovered from their pit burial. Somehow they didn't quite fit the picture. They *had* to be Neanderthal. But were they?

At a cave near Galilee, in Palestine, fragments of a female skull were collected. Although the bones were found in direct association with Mousterian artifacts—the type invariably associated with Neanderthal—the reconstructed skull appeared to exhibit certain modern features. It was an unsettling discovery.

Then, in 1931, a still more puzzling find emerged when caves on the western slope of Mount Carmel (Plate 23), also in Palestine, were opened. At Mugharet et-Tabūn (the "cave of the oven") a complete female skeleton was found, identical with the Neanderthal remains from Europe. But at the sister cave, Mugharet es Skhūl (the "cave of the kids"), ten skeletons came to light that would raise perplexing new questions.

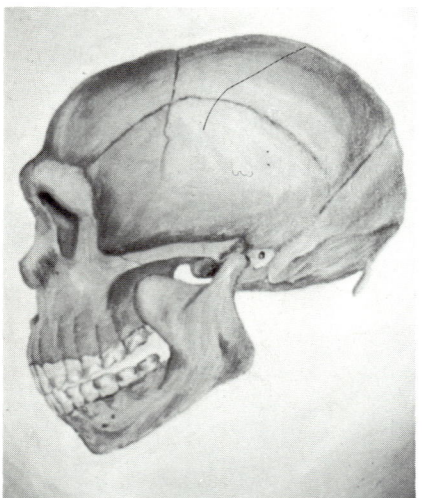 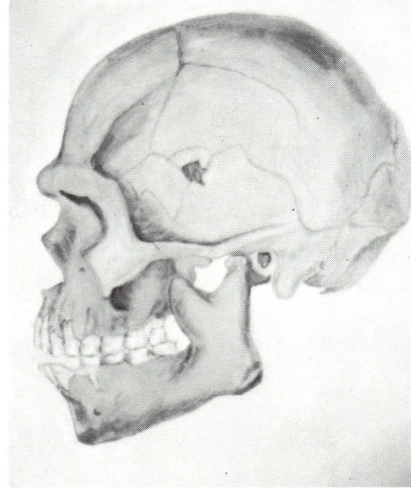

Plate 23. These drawings by artist Carol J. Swartz demonstrate the variability of Neanderthal man at Mount Carmel in Palestine. The skull from Tabūn Cave *(left)* is morphologically identical to that of the European Neanderthal. The skull from Skhūl *(right)* exhibits a higher skull vault, a smaller cranial capacity, and an incipient chin.

The Skhūl remains looked Neanderthal, but with a few startling differences. In the skull, the cranial vault seemed higher than usual for Neanderthal, and the projecting "bun" at the back of the skull, typical for European specimens, was missing. The skull itself seemed too small for Neanderthal: it measured only 1,400 cubic centimeters. And the mandible showed the slightest projection—an incipient chin, the feature lacking in all the Neanderthals found thus far. The post-cranial skeleton was almost modern in appearance: the limbs were neither thick nor bowed, and the legs were longer. In short, the Skhūl specimens represented a taller, more refined and gracile population than the European Neanderthals; they were neither wholly Neanderthal nor wholly *sapiens,* but a curious blend of the two. Each Skhūl skeleton combined ancient traits characteristic of Neanderthal with modern traits typical for present-day man.

The Skhūl discovery played havoc with existing theories of Neanderthal history. The Neanderthals so far uncovered had been remarkably homogeneous; that is, they were uniform in physical appearance, all unmistakably Neanderthal with no confusing combinations of primitive and advanced traits. Most archaeologists had assumed that with Neanderthal they encountered a local population which had suddenly and inexplicably become extinct, leaving no descendants to mix with the more modern types which, having evolved elsewhere, migrated into Europe to replace the Neanderthal populations.

But here at Skhūl, scientists examined almost a dozen skeletons that fit neither population with ease. Could it be that Neanderthal's extinction had come later than previously imagined, after interbreeding occurred with a modern population? Did the Mount Carmel remains represent a hybridization between the local Neanderthals and a yet-undiscovered ancient *Homo sapiens* race? Or—most exciting of all—had archaeologists at Skhūl stumbled upon a population caught in the very throes of evolutionary change?

Early theorists believed that the Mount Carmel remains represented a new type of Neanderthal, distinguished from the "classic" European group by their more modern features. The "classics," it was suggested, became extinct while the "progressives"—those of the type seen at Mount Carmel—evolved into modern man.

In view of subsequent discoveries, however, modern interpretations refute this early hypothesis. The early scholars badly under-estimated the Neanderthal populations in time and in space. The Neanderthals date not from 45,000 years ago, but from 100,000 years—or more. And they ranged over vast territories of the Old World. Isolated groups differed in certain physical features, par-ticularly in the shape of the skull.

Twentieth-century anthropologists have come to realize that variety is the rule rather than the exception. Just as we have racial variations in our present world, so must there have existed racial varieties of prehistoric peoples. It is a point overlooked again and again in scientific history. Poor Weidenreich was not alone in his later confusion!

Deserts, mountains, seas, lakes—these are simple barriers for man to scale today. But in those ancient days of Neanderthal man, geography effectively served to isolate small populations and severely restrict human migrations. And isolation favors the creation of racial differences both large and small. When breeding is limited to the members of a tiny population, that population tends to develop along its own unique lines. Denied the introduc-tion of new genetic material from outsiders, the members of an isolated group come to exhibit peculiarities in physical form and in culture that distinguish it from other isolated populations. This is how our modern races came into being. And this is how Neanderthal diversified.

And so we have come to know Neanderthal in his various guises. From Europe, we are acquainted with the "classic" Nean-derthal, distinguished by his low-domed skull, heavy-duty teeth, pronounced brow-ridges, and short, squat body. But we've had to revise our initial opinion of him. Recent studies prove that the European Neanderthal wasn't nearly so apelike as the early scien-tists imagined him. He walked as upright as any modern man. And certainly he was no less intelligent, although he lacked the rather dazzling technology we boast today.

"It is probable," says C. Loring Brace, anthropologist at the University of Michigan, "that if a properly clothed and shaved Neanderthal were to appear in a crowd of modern urban shoppers or commuters, he would strike the viewer as somewhat unusual in appearance—short, stocky, large of face—but nothing more than that. Certainly few would suspect he was their 'caveman' ancestor."

Modern anthropologists have dispelled early notions that the shape of the head correlates with intelligence or that the form of the brow denotes mental inadequacy. Even the presence of a forehead fails to indicate advanced intelligence, although we know now that Neanderthal man *did* have a forehead; his massive brow-ridges simply made it so low as to seem nonexistent.

From the Middle East, we recognize a second type of Neanderthal, the "progressive" variety. He more closely approximates modern man. He is taller and more refined; his skull isn't so low or long, and he has a higher, more visible forehead.

Neanderthals of every grade of variation in between have been found, and some authorities believe they have located Neanderthal in Africa. In Libya, archaeologists have uncovered the remains of a Neanderthal much like the type known from the caves at Palestine. At numerous sites in northwest Africa, we find traces of forms approximating the European Neanderthals.

And we find fossils that cannot yet be classified. A particularly mysterious skeleton comes from the depths of zinc mines located at Broken Hill, in northern Rhodesia, to astound anthropologists. The bones represent a male thirty to forty years old at the time of his death. He stood erect, on near-modern leg-bones, about 5 feet 10 inches tall. His skull—long and low and surprisingly narrow—resembles that of the classic Neanderthal, but the cranial capacity (about 1,200 cubic centimeters) falls short of the usual standard.

The bones look strong and robust, but Rhodesian man (Plate 24) isn't a picture of health by any means. His left leg-bone bears the marks of advanced rheumatoid arthritis. Ten of his fifteen teeth are badly decayed; a huge alveolar abscess presents a yawning cavity in the upper jaw. At the side of his head signs of severe infection, partially healed, mar the surface of the bone. A deep and ugly abscess triggered by acute ear infection has broken through the temporal bone itself.

Rhodesian man plucks at our imagination like few other fossil men. It is impossible to view his bony remains without being stirred at the thought of a form so rugged yet so mortally vulnerable to the advance of prehistoric disease. Scientists, spurning emotionalism, warn against reading too much into the skeletal record. Yet we can't help conjuring up a piteous picture of this ancient human form: that heavy head bowed with the pain of raging infection spread from jaw to ear; the massive brow throbbing with the agony of incessant headache; that huge, proud creature maddened and whimpering with fever and burning physical torment. Did he truly suffer so? Perhaps not. But the bones hint at more than mere body form; the most rational of scientists are caught by the fragmentary stories buried with individual bones. We search for fossil populations; often, we find the meager remains of men and women who lived and breathed and loved and suffered—just as we do today.

However much we are moved by the thought of Rhodesian man, he presents us with a few new problems to solve. He looks Neanderthal, but not quite. He looks *sapiens,* but not enough. What are we to do with him?

A few authorities judge him to be an early form of *Homo sapiens,* perhaps an ancestor to the present-day Australian aborigines. Others feel he better resembles Neanderthal man, but with several important differences; they place him in a separate species, *Homo rhodesiensis.* Many experts identify him as an African

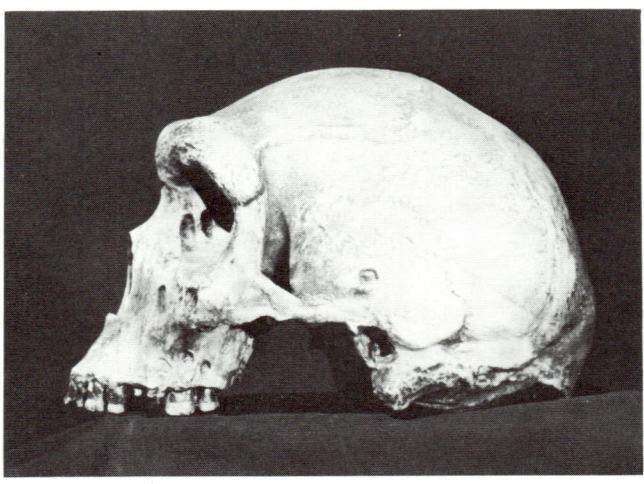

Plate 24. The skull of *Rhodesian man.* Note the hole in the temporal bone, believed to represent the effects of acute mastoiditis. *Photo courtesy Douglas H. Smith.*

variety of Neanderthal. And still others aren't certain; they beg the issue, calling him simply "Rhodesian man." At any rate, he is duplicated in many of his skeletal traits in another puzzling fossil from South Africa, near Saldanha Bay. Several authorities lump these two similar forms together for taxonomic purposes. There has been a recent trend toward authoritative acceptance of both the Saldanha and the Rhodesian remains as a subspecies of *Homo sapiens;* that is, *Homo sapiens rhodesiensis.* Does this mean Rhodesian man is not, after all, an African Neanderthal? The experts don't agree. You pays your money and takes your choice.

Another enigmatic fossil turned up in 1931, this time in Java, where the *Homo erectus* material lured armies of geologists and paleontologists. Members of a geological survey ordered by the East Netherlands government exposed in Ngandong deposits a puzzling skull and summoned G. H. R. von Koenigswald, a German paleontologist traveling nearby. Von Koenigswald, excited by the possibility of finding a Javanese Neanderthal, rushed to photograph the site—but in his exuberant agitation managed to overexpose his entire roll of film: we have not a single photograph of the original site.

No matter: von Koenigswald (who would later track down *Pithecanthropus II*) lingered to direct further excavations. Under his expert guidance, workers recovered eleven skulls, all without faces, over a ten-year period of digging.

Von Koenigswald, anxious to get on with his researches, began to feel the first tremors of war. Fearing the sudden arrival of Japanese troops, he quickly made casts of the skeletal material and parceled out the originals to his friends, asking them to bury the invaluable bones for safekeeping. Only the casts remained in von Koenigswald's laboratories: these were confiscated by invading troops. Von Koenigswald waited out the war in a concentration camp.

At the end of World War II, all of the original fossil material was recovered, thanks to the paleontologist's quick wit. Von Koenigswald and his fossils traveled to New York, where the bones were placed in the American Museum of Natural History. Here von Koenigswald sought the help of Franz Weidenreich (whom we met in the preceding chapter) in examining and interpreting the fossils.

For Weidenreich, it seemed unlikely that the fossils could represent a Javanese variety of Neanderthal. They dated from the early part of the Wurm I glaciation, a time contemporary with the European Neanderthals, but they seemed more primitive in form.

For Weidenreich, the whole find smelled a bit fishy. The human bones, he noted, consisted solely of skull-caps (Plate 25) and split long bones, although associated animal remains from the same deposit were complete and undamaged. He began to suspect that the bones had been artificially planted. Had someone salted the site?

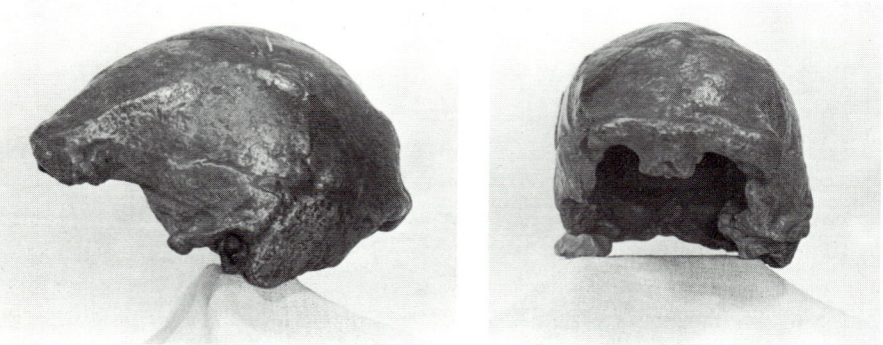

Plate 25. "Solo V," one of the cannibalized skulls from Java that represent *Homo erectus soloensis. Photo courtesy D. Gentry Steele.*

Then he looked more closely and concluded that Solo man was a cannibal—or the victim of one. Clearly, the skulls had been torn apart in an attempt to get at the brains; the long bones were split for their marrow. Here, after 50,000 years, were the shattered remains of yet another prehistoric cannibalistic feast! Like Black and Pei in China, Weidenreich had found irrefutable proof that violence is not a twentieth-century invention.

Who was the man-eating Solo man? It was a puzzle that would mystify Weidenreich to the end of his days. After extensive examination, he finally rejected the theory that Solo represents a Javanese Neanderthal, believing them to relate instead to the *Pithecanthropus* material—perhaps as a direct descendant of *Homo erectus.* But a final decision eluded him. His monograph describing Solo man was his last work; it ends in mid-sentence, and no one has ever completed it for him.

Solo remains a mystery. His skull is extraordinarily thick, resembling both *erectus* and Neanderthal. His average cranial capacity (about 1200 cubic centimeters) falls short of the latter, although his small brain size didn't keep him from making beautiful stone tools and bone axes.

But does he belong in this chapter? Some experts still consider Solo a Javanese Neanderthal. Others place him in a species of his own, *Homo soloensis,* while a few lump him into *Homo rhodesiensis.* But most authorities avoid making a taxonomic classification, dodging the question of identity by calling him, simply, "Solo man." No question, after all, is simple in the search for man's distant ancestors, and no answer is unqualified. This is a field in which experts do not hesitate to say "I don't know."

Still, the search goes on. At this very moment a young anthropologist at the University of Kansas, D. Gentry Steele, is working on a way to apply the principles of computer science to the puzzle of Solo's relationships. Steele knows that a man, however expert, can evaluate only a few skeletal traits at a time. The computer, however, can handle ten—or twenty or more—of these traits. Steele assigns numerical values to various skeletal features, feeds these into a computer, and compares the results with tests run on known fossils. Can this machine-age analysis give us Solo's true identity? No one knows. But if Steele succeeds, he will lead us into an exciting new research field pioneered by the eminent anthropologist W. W. Howells. If such men can obtain reliable results from computer programming, they'll add a valuable new technique to our present methods of identification.

We may not yet be able to recognize Neanderthal in all his various forms. But what do we do with the ones we've identified so far? Neanderthal leaves us with two primary questions: what became of him, and what was his role in the course of human evolution?

Early theorists believed that the Neanderthal populations were eradicated by an invasion of superior peoples, a sudden wave of *Homo sapiens* populations equipped with higher intelligence and advanced tools. These new peoples wiped out Neanderthal and took over his territory. Neanderthal became suddenly extinct, leaving no descendants to evolve into modern varieties of man.

Today, we know that Neanderthal was at least as intelligent as we are. We recognize the extent of physical variability among Neanderthal populations. And it seems unlikely to us now that *sapiens* vanquished Neanderthal in the usual sense of the word. More probably, Neanderthal man met early *Homo sapiens* and intermingled. Breeding took place between them—as always happens when human groups come into contact—and, gradually, the traits typical for later populations dominated in the offspring.

But what of Neanderthal's role in evolution toward modern man? When his remains were first uncovered, scientists overemphasized Neanderthal's primitive features. They were convinced that the Neanderthals were too different from us ever to have played an active part in our evolution. Those massive brow-ridges, the barrel chest, the heavily developed musculature combined to present specializations that could never melt away into the more generalized form of present-day peoples. But was Neanderthal truly so different from us that he fails to qualify as an ancestor?

Brace and other modern anthropologists believe that, given a suit of clothes and a manicure, Neanderthal man could pass unnoticed in our midst: that his differences fall well within the range of variation common for modern man. And if this is true, it is possible that we can trace our development through him.

The majority of experts hold to a refined version of the classic-progressive theory devised in the last century. This hypothesis holds that the "progressive" Neanderthals (those exhibiting a more generalized form) represent a phase in our development, while groups that were highly specialized (the "classics") did indeed become extinct. Since the latter group comes from a mutual ancestor, the classics are collateral cousins to modern man but played no active role in our evolutionary history.

Most authorities today realize that Neanderthal shares with us so many important similarities that he belongs in our species, *Homo sapiens*—but that he differs just enough to warrant a subspecies of his own. And so while we are classified *Homo sapiens sapiens,* he is known as *Homo sapiens neanderthalensis.*

Monsters, Races, and Modern Man

From the time of the early Greeks far into the nineteenth century, monster-hunting was a favorite pastime. No one doubted the existence of strange and eerie monsters: even educated men believed that the remote regions of the world were inhabited by fierce and exotic manlike creatures. Unicorns, mermaids, two-headed giants, winged horses, twelve-toed dwarfs, hairy ape-men, and fish-tailed gods ran rampant through the minds of men.

Museums and road-shows vied with one another to collect for display the most outlandish of forms taken from distant lands. Monster-chasing became so popular a pursuit that European governments learned to fatten their treasuries by imposing high taxes upon the importation of monsters. Wealthy collectors filled their curio cabinets with a vast array of odd forms, most of which owed their existence to the fertile imaginations and nimble fingers of skilled taxidermists. At a time when two-headed monsters were commonplace, it was little bother (and remarkably profitable) to sew a second head onto the body of a newborn calf.

In those early days when vast continents remained unknown, contact between isolated peoples was limited. Who knew what horrors lurked in the depths of a foreign jungle? Most people—ignorant of distant lands and unknowledgeable about the limits of human variation—populated the far corners of the earth with creatures of their own imagining. Travel, of course, was an enterprise reserved for a few highly imaginative and adventurous men. Bent on preserving their reputations as men of daring, they embellished their reports with wild tales of strange and ferocious subhumans. It would not do to disappoint a populace hungry for news of an heroic exploration.

Most intriguing were the beasts reported by early travelers. The Greek historian Herodotus, writing of his voyages in the fifth century before Christ, tells of encountering wild Egyptian peoples who lived underground and fed solely upon snakes and lizards. Ktesias, a Persian physician of the fourth century B.C., describes a race of Indians who had but a single foot on which they hopped faster and more efficiently than any hare. They made further use of this helpful appendage by employing it as an umbrella, holding it aloft to screen out the harsh rays of the sun and protect against rain. Medieval travelers reported hordes of five-eyed peoples, of men with ears so long that they dragged upon the ground, of weird manlike beasts with four arms and eyes in the back of their heads.

Early explorations into Africa and Asia brought Europeans face to face for the first time with the tropical apes, raising a new crop of hairy-monster tales. The conquest of the New World and the first sighting of American aborigines gave rise to wild stories of bloodthirsty and barbaric savages. European explorers, astonished and fearful, reported their encounters with imaginative embellishments, grossly exaggerating the oddities of habit and physiognomy of new-found peoples. The result was an immediate denial of human status to most of the inhabitants of newly discovered regions. Ordinary peoples and animals were transformed into beasts of extraordinary appearance; harmless natives became fierce man-monsters whose customs filled with horror the hearts of "civilized" men and women.

With so many strange man-beasts scattered across the world, there was little wonder that many of them made their ways into the earliest taxonomies. Even Linnaeus felt compelled to deal with these peculiar man-monsters. In his monumental eighteenth-century classification of life forms, he created a special species called *Homo monstrosus,* or "monster men." And into this fascinating category he lumped together all the fanciful creatures not entitled to inclusion in our own species. Not for another hundred years would it become clear that there exists but a single species of living man—*Homo sapiens.*

Man is represented today by more than three billion individuals spread over the earth's vast surface. He comes in various shapes, colors, and sizes. There may be as few as three or as many as 100 different human races, depending upon who is doing the counting and what physical traits he uses to distinguish between separate races.

Nevertheless, every living person belongs to the same species and the same subspecies. The mysterious Ainu of Japan, the tattooed Melanesian, the forest-lurking pygmy, the solemn Navaho, the Africans and Americans and Arabians and Vietnamese share a single taxonomic designation: *Homo sapiens sapiens,* separated from Neanderthal man (*Homo sapiens neanderthalensis*) only at the subspecies level.

We may track such whimsical modern monsters as Tibet's perplexing *Abominable Snowman* or California's puzzling *Big Foot.* But we may be assured that our world today is inhabited by only *one* genus, *one* species, and *one* subspecies of man.

For the archaeologists and paleontologists of Darwin's time, however, monsters remained very much in vogue. Man, dating his own creation at 4004 B.C., was not yet permitted a prehistory. He believed that there could exist no extinct forms of man, for extinction implies evolutionary failure. And, if each kind of plant and animal had been separately created—"ready-made" in its present form—no evolutionary process could have occurred. The species were immutable, fixed in their original forms by God at the instant of Creation. Any unknown man-forms from the distant past *must* then be monsters.

Then the brutish Neanderthal bones surfaced, eventually convincing all doubters that ancient man had indeed existed. Now science abandoned the "special creation" theory and began to stalk prehistoric humans, fully expecting that any fossils found would be at least as primitive and apelike as Neanderthal—or more so. Thus when the first representatives of modern man were found, they were ignored: they simply didn't differ radically enough from modern skeletons!

No one knows when the earliest forms of *Homo sapiens sapiens* appeared. But their bones were first revealed in France, in 1852. A French roadmender was hiking near Aurignac, in the south of France. It was his job to spot and repair any holes in the road—but his thoughts had strayed to the joys of rabbit-hunting. When he glimpsed a flash of soft white, visions of rabbit stew filled his head; his road-mending duties forgotten, he gave chase, cornering his quarry at a heap of large rocks a short distance from the road. Reaching into a narrow crevice after the wriggling hare, his fingers closed about a hard, linear object. To his horror, he found that he grasped not a rabbit at all, but a complete human leg-bone. The roadmender began to dig, shoving aside the heavy boulders. And by the end of the day he had unearthed seventeen human skeletons, with a cache of delicately worked tools of bone, flint, and ivory. Mystified, he summoned village officials who, after hurried consultation, decided that the skeletons belonged to modern man—perhaps they were the victims of a long-past mass murder. They treated these 30,000-year-old bones to a Christian burial in the village cemetery, never dreaming that what they had found were the remains of members of a population that followed and replaced the Neanderthals in most areas of the world.

In Europe, a number of scattered sites began to reveal the bony remains of these early moderns, and the immense rock shelters and caves of southern France produced the largest and most dazzling skeletal collections. Perhaps the most famous single site was that of Cro-Magnon, named after the cave-riddled region near Dordogne, France. Edouard Lartet, a successful attorney who gave up his lucrative law practice for the more challenging avocation of fossil hunter, suspected as early as 1868 that Cro-Magnon might prove a haven for the bones of early man. For several years Lartet scoured his own vast estate in search of ancient fossils. Now he turned to Cro-Magnon, and soon retrieved from Pleistocene layers five complete human skeletons together with advanced stone tools and extinct mammal bones. The human bones had been deliberately buried, clear evidence that these ancient French residents knew and practiced funeral rites.

Although the bones looked almost modern, there could be no doubt that they were very old: the geology in which they were found was unquestionable. Even so, numerous anthropologists led by the eminent Gabriel de Mortillet hotly disputed the suggestion that fossil man was sufficiently advanced to practice the veneration of the dead.

But the human skeletons from Cro-Magnon fell under elaborate scrutiny by E. T. Hamy and Armand de Quatrefages, respected authors of one of the earliest and most thorough descriptions of human physical types.* These learned men concluded that the bones represented the prototype of a new fossil race—the race of *Cro-Magnon.*

Their argument was bolstered by re-examination of a fossil from Paviland Cave in Wales. Found in 1823, this skeleton, called the "Red Lady of Paviland" because the bones were red (but actually represented an adult male), had been unearthed long before scientists were ready to admit the existence of prehistoric man. The "Red Lady" had been judged an intrusive burial, placed in the Oxford Museum, and forgotten. Now Hamy and de Quatrefages brought the bones out of storage, pointing out that this headless skeleton from Paviland demonstrated dozens of anatomical similarities to the remains from Cro-Magnon. When subsequent discoveries at Laugerie-Basse and Duruthy added additional representatives of this ancient fossil race, most contemporary authorities accepted both the geologic age (about 30,000 years before the present) and the racial status of Cro-Magnon.

Who was Cro-Magnon? Physically, the members of this first-named fossil race were tall and sturdy individuals. They stood between 5 feet 6 inches and 6 feet in height as estimated from recovered lower limb bones—taller than our average modern man. Their forearms and shins were relatively elongated. Their skulls were massive, surpassing a 1,600-cubic-centimeter cranial capacity. And their heads were long relative to breadth. At the same time, the face was broad and short, with a refined brow and distinct chin, giving the entire skull a disharmonious appearance still found among some peoples in northern Europe today.

Culturally (Plate 26), Cro-Magnon was a master craftsman. His implements of stone, smaller than those fashioned by Neanderthal, surpass all earlier stonework in diversity and workmanship. He

* *Crania Ethnica: Les crânes des races humaines* (Paris) 1882.

Plate 26. Reconstruction of the life of Cro–Magnon, from a diorama at the Museum of Natural History, the University of Kansas. *Photo courtesy Larry G. Quade.*

had turned to bone tools, too, producing from the bones of cold-weather fauna many lovely and useful implements.

But he was an artist and a sculptor too. He painted upon his cave walls, on bones and pebbles dramatic and realistic drawings of animals. Later caves are filled with magnificent polychrome paintings. And when he did not paint, he engraved, in meticulous detail, the forms he wanted to duplicate or sculpted in ivory and stone.

With the recognition of this fascinating fossil race, the scene was set for the identification of other races. In 1872, scientists began excavation of the nine caves at Grimaldi, located near Monaco on the Italian Riviera. Diggers recovered first the bones of "Mentone man," an adult male resting in his side with knees partially flexed. Additional specimens followed: three complete skeletons emerged in 1873 from the Cave of *Baousso da Torre* and, in successive years, two infant skeletons were taken from a nearby shelter, one known ever since as the *Grotte des Enfants*. All dated from late Pleistocene times. And all appeared remarkably modern, though found in association with shell ornaments and the bones of extinct animals.

The bones of most Grimaldi individuals were red. Presumably, the bodies were smeared at the time of death with red ochre in a funeral ritual. As the flesh decomposed and finally disappeared, the red pigment clung to the bones, giving the impression that they had been painted.

In 1901, anthropologists were summoned to a new site at the lowest level of the *Grotte des Enfants*. Within a single shallow grave, the bones of a thirty-year-old female and a sixteen-year-old boy lay side by side, their lower limbs flexed. A hollowed stone slab protected the skull of the young male; a bead head-dress and shell bracelets lay scattered nearby. Again, the bones gleamed eerily red in the morning light.

But this strange double burial added something new to fossil history. Investigators claimed to perceive in the bones unmistakable Negroid traits. Basing their theory on a clumsy reconstruction of the skulls and certain features of the pelvis, foot, and leg, they quickly established a new fossil race, that of *Grimaldi*, arguing that the paired skeletons who served as the only representatives of this race might well be ancestors to modern Negroid groups. Today, experts believe that the Grimaldi remains are not so black as they have been painted; they are reluctant to accept a Negroid status of these ancient bones on the basis of a hasty reconstruction.

The search went on for other fossil races. In 1889, two archaeologists from Périgueux recovered a single tightly flexed male skeleton from a rock shelter near Chancelade. The unnatural position of the skeleton (knees drawn up to meet the chin) suggested a deliberate burial. And the bones clearly resembled those of a modern Eskimo. Professor L. Testut immediately established another race—*Chancelade*—an ancestral race leading to modern Eskimo groups. Sir Arthur Keith later pointed out that this practice of creating new fossil races on the basis of single skeletons was recklessly extravagant. Besides, complained Keith, anyone could see that the Chancelade skeleton wasn't Eskimoid at all, but a true fossil European.

Nineteenth-century investigators, innocently expecting all fossil *sapiens* populations to be anatomically identical, cannot be blamed for going overboard in naming separate new races. They could not accept the wide range of physical variability encountered in late Pleistocene skeletons. And their naïve tendency to identify ancestral races on the basis of one or two specimens led to the persistent notion that "pure" races arose at the start of human history only to be "mongrelized" through mixture.

But what is race, after all, other than the simple adaptation of human populations to local conditions? Certainly mixture occurs, for, when people meet, they generally get around to mating. But "pure" races? Racial mixtures must have occurred over and over again from the earliest times; the only "pure" race is one which lives in absolute isolation from all other members of the human species, a circumstance that seems to have occurred very rarely in human history.

Modern anthropologists recognize that man is *polytypic;* that is, he comes in assorted types, which are generally referred to under the catch-all term *race*. That term has been used for certain objectionable purposes, and, to avoid becoming implicated in "racial" questions, some scientists prefer to use a term such as *biological unit* or *human population*. On the other hand, the term *race* is more familiar and may serve as well. *Race* may be scientifically defined, without opprobrium, as a *breeding population distinguished from other breeding populations in the frequency of certain hereditary traits.* In other words, the members of a given race tend to have in common numerous distinctive features —a particular color of skin, or form of eye, or dental structure. *Most* (but not all) Negroes have dark skin. *Most* (but not all)

American Indians have prominent cheek-bones. *Most* (but not all) Polynesians are tall.

There are "geographical races" (groups isolated by such major geographic barriers as oceans), "local races" (groups separated by distance, cultural habits, or lesser geographic barriers), and "micro-races" (groups marked by slighter regional differences). The last group may be scarcely distinguishable from other populations. In a large city, for example, such invisible racial groups may arise from the long continuance of peculiar settlement or mating patterns. And such patterns are reinforced by proximity: With millions of potential mates available, the typical male tends to choose one near at hand.

What causes races to develop? Certainly isolation is a factor, although it need not result from geographic barriers. Mating is selective, often guided by standards of beauty, and those preferred standards differ in each population. Among certain tribes in East Africa and in parts of Polynesia, fat women are considered the most desirable, while American women pursue slimness through endless dieting. Cultural regulations, called *customs*, help to enforce breeding patterns, too: among the present-day Trobriand Islanders, tribal males must select their brides only from neighboring villages.

Isolation and social selection both serve to preserve racial differences, but those differences arise initially through mutation and natural selection. A mutation is a spontaneous change in genetic material that can be transmitted to succeeding generations. Many scientists believe, for example, that light skin first appeared as a mutation. The earliest forms of man lived in hot, tropical regions where dark skin afforded ample protection against the sun, while permitting sufficient absorption, via the sun's rays, of Vitamin D, necessary for proper growth and development. As early groups migrated north, however, their need for such protection was diminished; in northern climates, the sun tends to be obscured by clouds. Light skin appearing as a mutation proved more effective in permitting needed absorption of Vitamin D, and thus light skin was propagated through natural selection.

Dramatic examples of natural selection abound in the animal world. Witness, for instance, the giraffe's long neck, which enables him to munch the tenderest leaves at tree-top level; or the wing-spots of the caligo butterfly, which permit the butterfly to startle his predators and give him a chance to escape. And natural selection operates on man just as it does among others of the animal kingdom. All the climatic forces we note about us are potential selective agents bringing about, through adaptation and mutation, the differentiation among races. A fascinating example of natural selection in man is seen in sickle-cell anemia, an often fatal condition that causes the destruction of red blood cells. Early investigators noted a high incidence of this disease among Negroid populations, particularly those living in areas where malaria is common. Since the disease is normally fatal, scientists suspected that its high incidence must have some selective advantage. And they were right: investigation proved that sickling is inherited in *two* forms: the mild sickle-cell *trait* and the crippling sickle-cell disease. If an individual inherited the gene for sickling from both parents, he would experience the blood-destructive disease, and his chances for death at an early age would be extremely high. If, however, he inherited the gene from only *one* parent, he would carry the trait but escape the disease. Persons who carry the mild sickle-cell trait are more resistant to malaria than are normal individuals. They do develop malaria, but less often and to a milder degree. Because malaria is also a dangerous condition, evolution favors the retention of the sickle-cell gene—a natural protection against death from malaria—in the breeding population. Man is caught in a game of evolutionary roulette, involuntarily and unknowingly risking death by anemia in exchange for a healthy immunity to malaria. Such a phenomenon is called "double selection."

Although we vehemently reject the early haphazard racial designations made on the basis of meager fossil evidence, we have not yet learned all there is to know about race. Nor have we managed to free ourselves entirely from the racial commitments contracted by those early investigators. These linger on in our textbooks, persistent hangovers from a scientifically naïve past. It is probable that racial differentiation as we know it today had its beginnings at various times during the Pleistocene. But the fossil evidence gives us no indication of when—or where—such differentiation took place.

The archaeologists of the nineteenth century had no appreciation of the extent of human variability, and certainly the wide range of physical differences exhibited in the early *sapiens* fossils puzzled them. Their solution—perhaps the only predictable one for that time in scientific history—was to designate racial types on the basis of polytypic fossils. But it was only a matter of time until they would come up against a population for which they had no racial category.

One of the most bizarre fossil discoveries came at Predmost, in Yugoslavia. Predmost had long been known as the site of an ancient mammoth cemetery; man-made tools and human skeletons had occasionally been found mixed among the mammal bones. But in 1894, workers shoveled aside a new floor of rubble to find a tight circle of some forty complete human skeletons, each in squatting position and protected by a crude stone rampart. Bewildered excavators finally decided that they had stumbled upon some sort of mass ceremonial burial. Astonished as they were in finding the remains of a mass ritualistic funeral, the tantalizing thing about the site was the fact that Predmost man (Plate 27) failed to fit any existing racial classification.

Nor did he appear to represent a new subspecies or a new race. Instead, each skeleton seemed a curious mosaic of racial traits, combining Neanderthal features with those typical for Cro-Magnon, with modern Australoid and Negroid characteristics thrown in for good measure. The investigators came to the conclusion that the Predmost skeletons were intermediate between

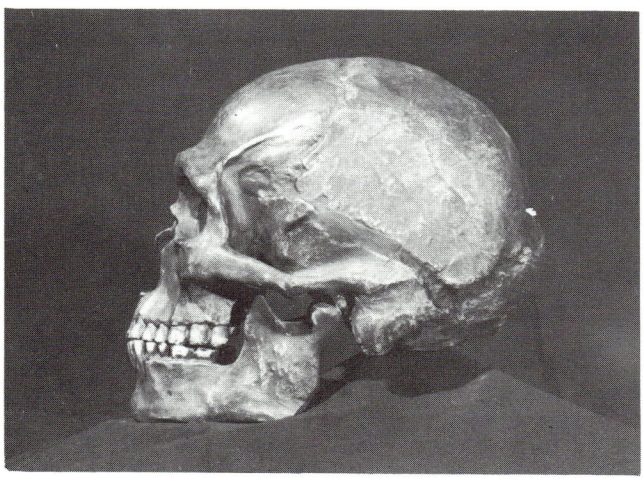

Plate 27. Cast of reconstructed skull from Predmost, Czecho-slovakia. *Homo sapiens sapiens. Photo courtesy Douglas H. Smith.*

Neanderthal and Cro-Magnon. The people of prehistoric Pred-most represented a hybridization of Neanderthal and some early Cro-Magnon group.

This is an interesting idea, for it acknowledges for the very first time the possibility that Neanderthal and *sapiens* populations lived side by side during the same time period; and, further, that these two groups interbred, giving rise to modern European populations. For nineteenth-century science, this was a revolutionary concept.

But our point is this: Predmost produced a skeletal sample that was different from any others so far unearthed. And it represented a variability that demanded explanation—not as an example of the normal variability of Europeans some 30,000 years in the past but as something special: something that must be accounted for because it was so totally unexpected.

Now new fossil types appeared with astounding frequency. In 1928, Professor Camille Arambourg recovered more than fifty complete late Pleistocene skeletons from a cave at Afalou-Bou-Rhummel in Algeria. Similar skeletons emerged from other sites in North Africa. Collectively known as the *Mechta* type, all of these remains resembled the European Cro-Magnons. The Mechtas were tall, with broad sloping shoulders and elongated limbs. Their skulls were large (cranial capacity well over 1,500 cubic centimeters), strikingly coarse and brutish in appearance. Like Cro-Magnon, they combined a long narrow skull with a short, broad face. Only the nose differed: for Mechta, the nasal aperture was wider.

Some authorities perceive Neanderthal as well as Negroid traits in the Mechta remains. Neither are easily recognizable. At any rate, it is not surprising that any African skeleton receives special scrutiny for Negroid traits. What *is* amazing is that not one single skeleton from the African Pleistocene can be portrayed as typically Negroid and similar to those groups found today south of the Sahara. But one cultural feature does tend to bolster the theory of a Negroid ancestry for Mechta: Each skeleton exhibits the marks of artificial mutilation of the front teeth. The two upper middle incisors had been extracted before death, a practice today associated with coming-of-age ceremonies among certain African peoples.

The same dental mutilation is noted in another provocative

fossil from the desert sands of the western Sahara. At Asselar, a
military post a few miles northeast of Timbuktu, excavators recovered a late Pleistocene skeleton representing a mature male.
The skeleton seemed to combine Mechta traits with Negroid
features. And some exuberant experts decided to relate the Asselar
remains with the Grimaldi skeletons of Italy as well as the modern
Bantu and Hottentot populations of South Africa. But this is an
ambitious hypothesis to base upon a single lonely skeleton.

From such sites as Boskop, Tzitzikama, Fish Hoek, and Spring-
bok—all in the Transvaal of South Africa—come a number of
skeletons known collectively as the *Boskop* type. With their
amazingly large skulls (some boast cranial capacities of 1,700 cubic
centimeters) and small faces, many authorities make them out to
be late Pleistocene ancestors of the modern South African Bush-
men.

But Boskop was not the only late Pleistocene form to roam
South Africa. From Florisbad, in 1933, came a single puzzling
skull found in association with the bones of extinct animals and
stone tools that seem remarkably similar to those made by Neanderthal in Europe. The Florisbad skull is large and rugged,
resembling both Neanderthal and Rhodesian man—but in many
ways it is more advanced than either. Like the Predmost population from Yugoslavia, Florisbad represents a transitional type between Neanderthal and *sapiens*. And since it predates Boskop in
Africa, many experts believe Florisbad is ancestral to all the
Boskopoids.

Early *sapiens* fossils abound in both Europe and Africa, but
they're not so easily found in other areas of the Old World. Expeditions into the Middle East bent on finding the earliest *sapiens*
representatives have had disappointing results. In 1938, two American Jesuit priests excavated a cave at Ksar Akil, about seven miles
northeast of Beirut in Lebanon. Fathers J. G. Doherty and J.
Franklin Ewing exposed a single skeleton, that of a male who had
died at the age of seven. These bones were promptly nicknamed
"Egbert," a name the young skeleton has never been able to live
down. Egbert had a perfectly modern braincase with no brow-
ridges, a high forehead, and firm chin. And that's just about all
we know about Egbert, except that he lived during the Wurm
glacial advance.

The skeletons of five adults and one infant were found in a cave known as *Djebel Kafzeh,* near Nazareth in Palestine in 1934 and 1935. They were physically similar to the Cro-Magnons of Europe and lived about the same time as Egbert.

From Russia come many exciting cultural artifacts from the late Pleistocene, including beautifully carved figurines in ivory and stone, usually representing a female figure with swelled breasts and buttocks; these are believed to be fertility symbols meant to celebrate the reproductive force in nature. Bracelets and other ornaments are plentiful here. But human skeletal remains are surprisingly absent. One of the few exceptions comes from Malta, in Siberia; it is the skeleton of a child, colored red and decked with delicate ornaments.

But the Far East has produced more interesting skeletal evidence. We have already discussed the brutalized family discovered at the Upper Caves of Choukoutien—the skeletons that seemed to Weidenreich to suggest racial mixture. In fact, racial diagnoses are attached to almost all of the skeletal finds from this area. Skeletons that appear to be Mongoloid—or proto-Mongoloid—emerge from such sites as Choukoutien, Tazeyang, and Liukiang, all from China's late Pleistocene. Further south, the two Wadjak skulls from Java have been classified as proto-Australoid, or ancestral to the modern Australian aborigines. The same category is acceptable for the Keilor skeleton from Australia. And from northern Borneo's immense Niah Cave comes the skull of a young female buried with advanced stone tools very similar to those from other sites. She dates from about 40,000 years ago and strongly resembles the native peoples of Tasmania.

All of these skeletons are representatives of our subspecies, *Homo sapiens sapiens* (Plate 28). All are noted for certain characteristics that make their owners likely ancestors of the world's modern races. Certainly the peoples of late Pleistocene times gave rise to our modern groups. And it's interesting to puzzle over their possible racial kinships. But the identification of race in modern man is often difficult. This matter of identifying race in a 30,000-year-old skeleton is particularly risky.

All we really know is that the Neanderthals disappeared from the fossil record to be replaced by numerous diverse types of modern-looking peoples. Apparently, there was no single point of origin from which this new modern type spread suddenly across

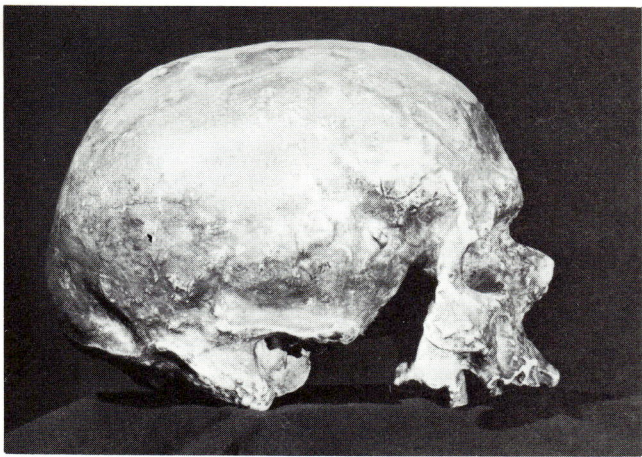

Plate 28. Cast of skull of early *Homo sapiens sapiens,* previously called Cro–Magnon. *Photo courtesy Douglas H. Smith.*

the world; it is more likely that far-flung populations reached this stage of evolutionary progress at roughly the same time—between 30,000 and 40,000 years ago. Physically, they are very much like modern peoples. Culturally, they had made great technological strides over the Neanderthals. They had learned to refine the sort of stone tools favored by Neanderthals, now making them smaller and more efficient. They invented a great variety of new tool types (Figure 7), many of them specialized to fulfill particular purposes. They had gained considerable skill in manufacturing stone tools, but preferred instead to work in bone, ivory, and antler. An important advance was the refinement of the *burin,* a pointed cutting tool of wide utility. Flint knives are seldom strong enough to cut bone efficiently. But with the burin, engraved bone implements become abundant. In this way, the peoples of the late Pleistocene were able to replace hardwood, which is lacking in many regions.

These early moderns required specialized and refined tools, for they were skilled hunters whose survival depended upon the efficient exploitation of a particular species of game—horse, mammoth, or reindeer. These were slaughtered not only for food: raw materials such as bone, sinew, skin, and antler were taken for clothing, wind-breaks, and tool production. Projectile weapons appeared, suggestive evidence that these early groups had learned

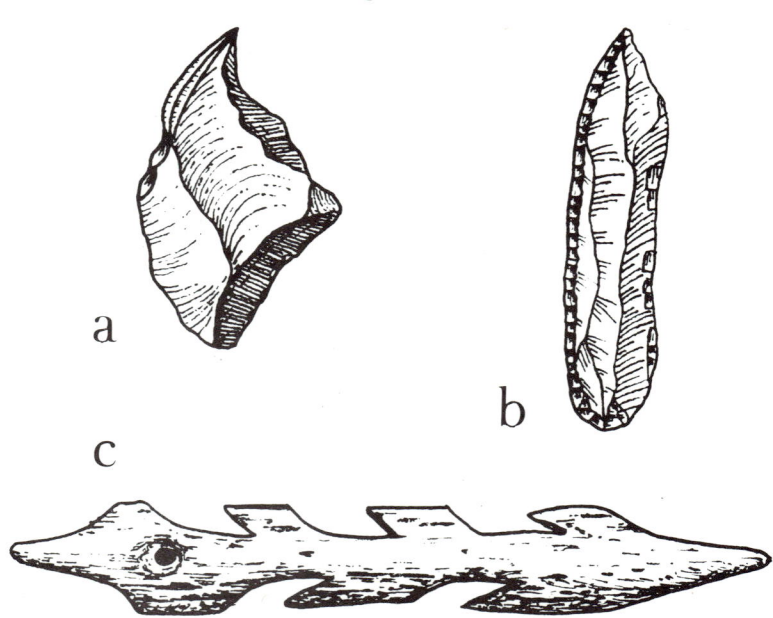

Figure 7. Typical tools of the late Pleistocene (Upper Paleolithic period): (a) Aurignacian burin, (b) Aurignacian backed blade, (c) Magdalenian bone harpoon.

to use artificial propulsive devices to increase the leverage of the human arm. No spear-throwers have yet been found, but it is likely that they were used, together with spears, barbed harpoons, and arrows—all implements usually hafted onto a detachable shaft. Eventually, most stone tools were replaced by implements of bone. Bone offers several advantages over stone: it is stronger and more flexible. It can be fashioned into such small tools as fish-hooks, harpoons, and needles.

True eyed needles are found in the later sites, proof that the early moderns wore clothing. Ivory beads and bracelets of ivory, shell, and teeth substantiate this theory. Adornment was an important part of dress in late Pleistocene times.

These early groups of *Homo sapiens sapiens* were our first artists and sculptors. They painted upon their cave walls—drawings in faint outline, single-colored flat wash sketches, and magnificent polychrome drawings. Usually, theirs was animal art, representing in stylized form the animals important to their survival. They seldom depicted scenes of nature or human forms, for they were preoccupied with the bison, horse, deer, musk-ox, ibex, mammoth, rhino, bear, and reindeer about which their lives revolved. Smaller food animals and inedible predators appear rarely. On pebbles, knives, and other implements these ancient artists inscribed similar forms. They left behind carvings, too, and low-relief sculptures. Perhaps their most distinctive art objects were the female figurines today called "Venuses." These, carved in ivory or modeled in clay, depict lush, full-bosomed females, many of them obviously pregnant. Were they prehistoric pin-ups—or early fertility symbols? No one knows for certain. But we suspect that the intriguing Venuses symbolize for the early *sapiens* groups the crucial reproductive force in nature.

Most late Pleistocene art is hidden in the inner reaches of caves, often painted over three or more times; apparently, the paintings served magic purposes. Perhaps it was a practice before the hunt to sketch the slaying of food animals; this is hunting magic, designed by men who believed in gods to better their chances in the hunt. Whatever its purpose, this artwork is unexcelled. Paintings preserved deep within France's Lascaux Cave are so splendid that Lascaux is often called the "prehistoric Sistine Chapel."

By 40,000 years ago, early representatives of our subspecies had wandered into and inhabited all of the continental masses and most of the larger islands from England to Japan, from Siberia to southern Africa. There is nothing primitive about these peoples: they represent a huge advance over Neanderthal man.

But they knew nothing of agriculture or of the domestication of animals. They had never raised a city nor learned to write. Theirs was still a stone age, and they lived as hunters and gatherers, totally dependent upon their wit and skill at ferreting out sufficient food supplies to stave off starvation.

We have said nothing of the New World, for man did not evolve here. No ancient apelike fossils have ever been found in the New World. No primitive representatives of man ever reached these shores. Neanderthal never saw the Western Hemisphere. But at some time between 20,000 and 38,000 years ago, splinter groups began to cross into the New World via the Bering Strait. These were not the makers of that magnificent European cave art, not the possessors of the fine, delicately flaked tools—but loosely organized bands of roaming hunters, marginal peoples who approached the New World before the invention of the bow and arrow. They came from ice-bound Siberia, making their way into arctic Alaska on the trail of grazing herds—musk-ox, caribou, bison, woolly mammoth. These groups carried with them no more than a few crude flaked points and spears. Their goal was not exploration; they had no wish to conquer a new land—indeed, they were unaware that a new land existed beyond the icy Bering Strait. They simply walked blindly in the glacial mists, stalking the prehistoric mammals on which they fed.

The Bering Strait was easily accessible in those days. Even today, Siberia and Alaska are but fifty-six miles apart, and unbroken water ranges only half that distance. At the time of the original migrations, extensive glaciation had lowered existing sea levels so that the two continents were connected by a land bridge perhaps 1,000 to 1,300 miles wide.

At no time did there occur a single mass migration. But family units and small hunting bands trickled in over a span of thousands of years. Old men died along the way, new babies were born. Generation succeeded generation before the journey was completed. A few groups leaked back into Siberia, unaware that they had walked upon the face of a vast new continent. Others, funneled in through Alaska, eventually came upon the Yukon and Mackenzie rivers where fresh water and big game lured them ever onward (Figure 8). The immense ice-free corridor of the Yukon beckoned, carrying the grazing herds and the men who sought to eat them deep into the grasslands. From here, the splintered groups fanned out in almost every direction, a few settling in what is now California, others heading east. Some kept moving southward. Eventually, small groups of stone-age men trickled across both continents of the New World, even to the very tip of South America—to Tierra del Fuego.

There seemed for many years to be little interest among anthropologists concerning man in the Americas. A few experts took the position that there was simply nothing old in the New World, and concentrated their researches in the Old. But many others were determined to locate the ancestors of our American Indians. They searched for more than three decades with little to show for their efforts. Then, in 1927, J. D. Figgins of the Denver Museum of

Figure 8. Major migration routes into the New World.

Natural History came upon a single projectile point that would change the course of American anthropological history. It was a *Folsom* point, distinctive because of its longitudinal channels—called *fluting*—on each face. What excited Figgins was that the point was deeply imbedded between the ribs of an extinct variety of bison (Plate 29). Obviously, the point lodged there at the hands of a lost and ancient hunter. Figgins' find destroyed the thirty-year opposition to the theory that early man had roamed North America. Since 1927, Folsom points have appeared in sites in many parts of North America; these date from 9,000 to 10,000 years ago.

The bones of Folsom man have never been found. But his existence is accepted on the basis of this cultural evidence. Then came pre-Folsom material: at Sandia Cave, in New Mexico, roughly-flaked points have been dated at not less than 25,000

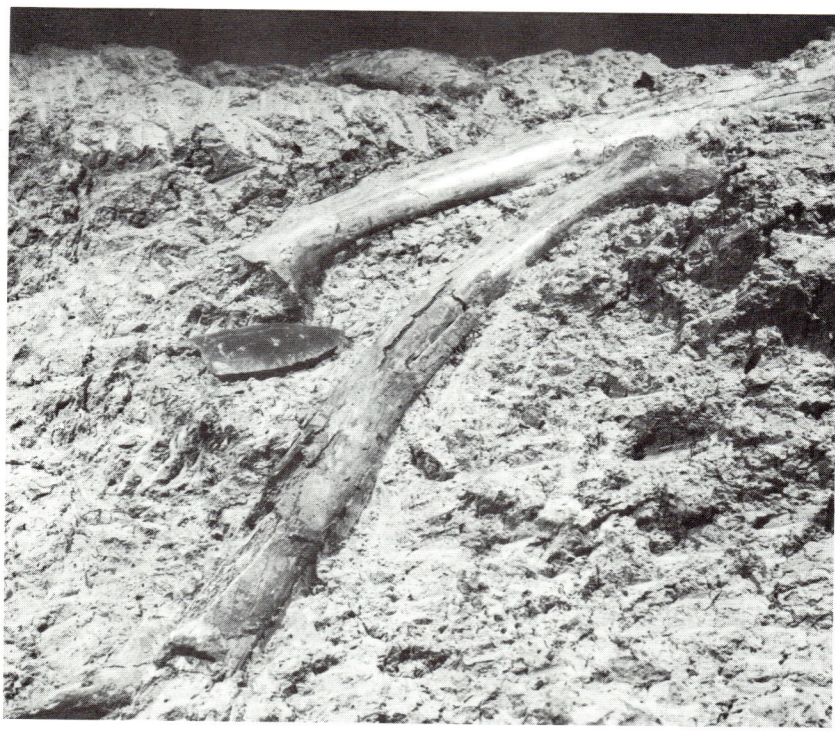

Plate 29. This Folsom point, photographed as it was found between the ribs of an extinct bison, proved for the first time the presence of early man in the New World. *Photo by Robert R. Wright, courtesy Denver Museum of Natural History.*

years ago. At Tule Springs, Nevada, the burned bones of camel and mammoth were found in association with man-made artifacts; this collection is dated by carbon-14 at more than 23,800 years. A site on Santa Rosa Island, some forty-five miles off the coast of Southern California, represents a huge mammoth kill: a concentration of burned bones and a single chipped stone implement give testimony to the fact that, nearly 30,000 years ago, America's earliest residents gathered here to feast on slaughtered mammoth.

Still, skeletal remains are rare. Many fossilized skulls and other bony portions to be sure have been offered over the years as our most ancient New World inhabitants. All have undergone the meticulous analysis of our best scientific minds; most have failed the rigorous geological and paleontological requirements for validity. A few have survived—and these few must provide us with our insights into the nature of these early Americans.

From a clay deposit in a deep ravine near Natchez, Mississippi, comes a hip bone with a long and complicated history. This pelvic bone was recovered in association with the bones of ground-sloth, horse, mastodon, and bison—all animals that had become extinct many thousands of years ago in America. Early time estimates were not much better than guesses, since the geological scheme for North America was not well known at the time of this discovery. The first estimate, given by the highly esteemed English geologist Charles Lyell, dated the deposit at 100,000 years old—a date we know to be impossible. Next, an early fluorine analysis made in 1895 showed that the human hip bone was older than the ground-sloth bones, implying that the associations were unsound.

In 1948, Professor M. F. Ashley Montagu revived interest in the Natchez material when he carried portions of the bone to England for new fluorine testing. The results showed again that the hip bone predated the bones of the ground-sloth. Not until later did we discover that the ground-sloth met extinction only 8,000 years ago: It was possible, after all, that the human hip bone had been buried before that date.

In 1953, wind and erosion exposed near Midland, Texas, a fragmentary human braincase. Once reassembled, the bones represented the top and back of a skull of what is probably a female aged about thirty years at the time of her death. Projectile points, grinding stones, and an incised bone were found with her. But

controversy rages over the Midland discovery: test results range
from 10,000 to 20,000 years. Because the bones were found in a
"blow-out" site, no geological evidence was present to substantiate
either date. Once more, we find human skeletal remains that cannot yield sound information concerning America's earliest natives.

Our first good opportunity to investigate the physical nature of
these early mammoth-hunters came in 1947, when a complete
skeleton was unearthed near the tiny village of Tepexpan, not far
from Mexico City. Using an electrical detecting device, Dr. Hel-
mut de Terra, an American geologist, and Dr. Hans Lundberg, a
Canadian, found the bones at a depth of four feet, lying in the
undisturbed layers of an ancient lake near the butchered remains
of two extinct mammoths. The skeleton (Plate 30), on the basis
of the geology of the site, was determined to be about 11,000
years old—a date substantiated by carbon-14 tests. The bones represent a male nearly sixty years old at the time of his death. He
differed in no way from the inhabitants of the same region today.

Sites in South America too have lured those who search for
America's oldest human bones. But most of the fossils found have
not withstood the test of time nor the scrutiny of modern scientists. Two sites, however, remain of interest to us.

In 1937, Dr. Junius Bird excavated a cave site, *Palli Aike,* in
Pategonia. He found numerous cultural artifacts, the bones of
extinct ground-sloth and horse, and the fragmented remains of
early man. The cave was dated by the carbon-14 method at ap-
proximately 9,000 years.

Located in the Lagoa Santa region of Brazil is a cave known as
Lapa de Confins. It was here that, in 1935, a human skeleton was
found under conditions that suggest great antiquity. No absolute
date can be derived from the remains, but the evidence points to
a time at least equivalent to the Palli Aike site. The Confins man
looks exactly as we would expect: in skull features and body form,
he is typically Amerindian.

By 10,000 B.C., then, primitive Asiatic hunters had crossed into
the New World on foot, spread, and inhabited most of North
and South America. They looked very much like their descendants, the Indians who populated these regions at the time of
European contact in the fifteenth century. There seems no doubt
that they originated in Asia: their physical characteristics and

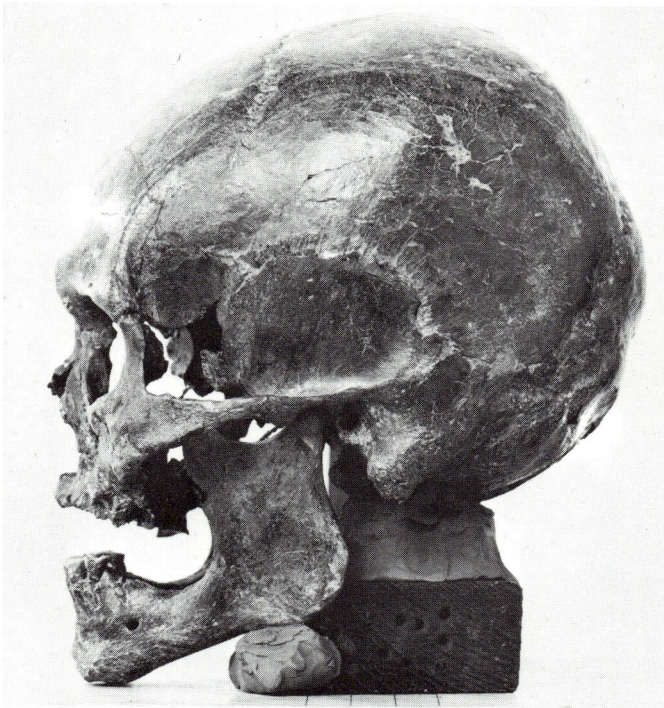

Plate 30. The skull of Tepexpan man, found near Mexico City. *Photo courtesy The Smithsonian Institution.*

their cultural artifacts are almost identical with those found in Siberia for the same time period.

With the end of glaciation, sea levels rose to obscure the path of their journey into the North American continent. They were cut off from the cultural developments of the Old World; they developed their own life ways. In time, they would give rise to the many diverse Indian tribes of North America; to the mysterious Mayan temple-builders of Yucatan; to the mighty Toltecs and warlike Aztecs of Mexico; and to the Incas and pre-Incas whose wealth and cultural achievements would amaze the Spanish *conquistadores,* who would come in ships to the New World in their search for riches.

But those are historic times. And we are not yet done with prehistory.

8

Mysteries in Prehistory

More than a century has passed since modern men first began to search in earnest for the fossilized remains of their most distant ancestors. We've been extraordinarily lucky in the recovery of many important fossils from the remote past.

Each has a story to tell.

Each draws a sketch of man on his way to becoming human.

Each contributes a clue to the story of man's evolution. Taken together and arranged in temporal sequence, these fossils enable us to trace in broadest outline the dramatic course of human evolution.

Man's origins lay in the late Mesozoic, when some yet unknown proto-primate, a tree-dwelling mammal of generalized form and vegetarian appetite, began to exhibit the traits that would lead to unprecedented evolutionary success. His brain grew larger and more complex. His vision became acute. His sense of smell reduced and his snout shortened. His front paws were transformed into grasping hands. His forelimbs swung free, permitting arm movement in almost every direction. Enhanced muscular coordination and a refined nervous system bettered his chances for survival. Life was secure at tree-top level: here ground-dwelling predators presented little threat, and food, available in arm's reach, was plentiful.

Perhaps such dazzling evolutionary success favored the development of a larger, heavier body for one or more of our distant relatives, making life in the trees less comfortable.

Perhaps it led to overpopulation, with an intense competition for food in the forests.

Or perhaps this generalized primate ancestor of ours was simply an inquisitive and restless animal.

But for whatever reason, he soon ventured down from the trees to feed upon roots and soft grasses. Abundant seeds and tender green shoots led him further and further into the grasslands, out into the open savanna. Somewhere along the way—for reasons we can only guess at—he became bored with his vegetarian diet and acquired a taste for red meat.

Now he was *Australopithecus.*

He lived, we think, much as a free-ranging baboon does today. He slept in trees or caves or on cliff ledges. He scavenged for food, robbing the kills made by swifter predators. Later, he pursued smaller game with his bare hands, bagging an occasional rabbit or picking off a young or weakened antelope.

And then he learned the secret that would make him man: Stones and branches, fashioned into crude tools, made powerful offensive and defensive weapons. So enlightened, he roamed with club or digging stick in hand, banded always in small troops. We visualize him in roving bands, rather like the modern-day wild baboons, with adult males in the lead and forming a protective flank for the females and children clustered in the group's center. *Australopithecus* was a man-in-the-making, quickly aware of the fact that there is safety in numbers. He was learning the advantages of social control and cooperation.

There was no need now to remain a little-league hunter: armed with tools and weapons, man could enter the major leagues, entering into direct competition with the predators of the savanna. The tribe was born: Men began to plan with other men the strategy of the hunt and to share with others, left at home, the game they killed.

More than half a million years ago—or perhaps as long ago as a million years—man broke out of Africa to explore a larger world. Quickly, he made his way to China. He tamed fire for his own uses. Now he could cook his meat, fend off the cold of winter, and drive away with blazing logs the giant carnivores that tried to stalk him. Equipped with these new advantages, he spread quickly to other continents.

Now he was *Homo erectus*.

He continued to slaughter smaller game when the opportunity arose. But now he was a more ambitious hunter. He located herd animals, driving them with sticks and branches, into swamps and mud-bogs where they would be helplessly mired until he slaughtered them. He learned to set deliberate grass-fires for the same purpose: to drive giant herds off cliffs or into the swamps where he could reap the harvest of a massive kill. Now he learned to talk, to plan in groups the complex strategy of bigger, more successful hunts and to reminisce about past hunting triumphs. His nomadic life was seasonal: he banded together with other primitive hunters when game was plentiful, then turned to the gathering of roots and berries when game was scarce. He began to divide his labor, taking the hunter's role for himself, leaving the women at home to keep the fires burning.

A hundred thousand years passed. Life was still precarious. Man pitted his wits against the rigors of a world filled with ice and dangerous carnivores. There was much he had not yet learned. And to keep his fears at bay, he found religion. He began to recognize death and to hope for an afterlife. He smeared the bodies of his fallen kin with ritual red ochre and buried them with offerings of food and implements and wildflowers.

For now he was Neanderthal.

His weapons were smaller, surer, more efficiently chipped from stone to produce tools for numerous diverse purposes. His social life improved. He learned the pleasures of companionship, of seasonal feasting and storytelling around a glowing fire, of magic designed to help assure survival. He began to care, as best he could, for his aged relatives. For his children, he provided a longer period of dependency. Since culture replaces instinct, a growing child must learn to be a man, must have time to master the complex techniques of weapon-making and fire-building and the intricacies of the hunt.

Then, suddenly—for a millennium is but a second in evolutionary history—Neanderthal was gone, vanished from his icy world to be replaced across the earth by men whose bodies differ little from our own.

Now man was *Homo sapiens sapiens.*

He had become a skilled hunter, abandoning the crude hit-or-miss methods of the past to concentrate all his energies on the full exploitation of a single species of herd animal. Whether he chose reindeer, musk-ox, or mammoth, he took from this one species not only the meat he needed to fill his stomach but also hides, horns, and sinew—materials he could transform into semi-fitted clothing and accessory implements and warm skin shelters. He disdained the crude stone hand-axes favored by his predecessors, making instead fine bone spears and harpoons. He invented gravers and burins, chisel-like flaked tools for working in bone, wood, ivory, soft stone, and antler. He discovered the joy of beauty for its own sake. His tools were now engraved with artful, delicate etchings. From bits of bone and shell, his deft hands fashioned fine bracelets, necklaces, beads, pins, and pendants to be worn by both men and women. He used cosmetics, applied to the face and body in life and in funeral ceremony. He decorated his clothing with colored beads and pierced teeth. And he made charms and amulets, objects designed to ward off evil spirits through their inherent magical properties.

His art was mixed with magic. The magnificent polychrome paintings and carvings that covered man's cave walls did double-duty: they were undeniably works of art—but they represent too man's attempts to deal with the supernatural. We call it imitative magic: to better one's chances in the hunt, one makes a picture of the animal to be killed, then sketches in the spear that will fell it and the blood that will flow from its wounds. Success will follow, hopefully, in reality.

Homo sapiens now ranged free over the world, venturing even into the New World. He had domesticated the dog, his constant companion. Soon he would abandon the free life of the hunter to settle in sedentary villages. He would domesticate plants and other animals. He would learn to work in metals. He would found the earth's first cities. He stood now at the very brink of civilization.

Prehistory ends (and history begins) not with the first appearance of a fully modern human type but with the invention of the written word—an event that occurred at different times in different parts of the world. Once man learned to write, he could leave behind him the story of his passing. His saga then becomes one for the historian to ponder, although it remains to the archaeologist to recover, through excavation, the crumbling fragments of the most ancient written records.

The end of prehistory is signaled in the Old World by the emergence of Sumerian civilization. About 4000 B.C., primitive farmers began to settle in the fertile Tigris-Euphrates Valley. Over a span of several millennia, their simple farming villages gave way to towns and cities, centers of religious and political activity. Our earliest evidence of true writing comes in the form of clay tablets heavily inscribed with pictograms. Many of these, dating from 3000 B.C., have been partially translated to hint at everyday life in this most ancient of all known civilizations.

History comes later to the New World. It was not until about A.D. 500 that true cities flourished here. Teotihuacán, the short-lived but influential ceremonial center in the Valley of Mexico, by now supported a dense population, probably through large-scale agriculture. In Quatemala, the mysterious Maya had reached their cultural peak, attaining dizzying heights in sculpture, architecture, astronomy, mathematics, and hieroglyphic writing. In Peru and Bolivia, various peoples worked skillfully in gold and silver, produced large quantities of fine textiles and pottery, and raised mighty armies. The Inca dynasty would not be founded until about A.D. 1200—but then the peoples of Peru would rival the ancient Egyptians in the splendor of their cultural achievements. They would bring under cultivation vast regions of arid desert land by means of intricate irrigation systems, span the continent with a complex network of paved roads, perfect the art of human mummification, and unite many diverse peoples and customs in building a mighty new empire. Amazingly, they would do so without ever developing writing. For practical purposes, then, we date history in the New World from the time of the Spanish conquest—from the sixteenth century.

And so to know prehistoric man, we must begin with those remote Australopithecines and sift through time almost into the present, depending upon where we dig. Perhaps our most exciting discoveries are those that add new dimensions to our knowledge of past groups. We are as delighted to find new specimens of known fossil populations as we are to uncover previously unknown groups. Each new discovery helps to link the known stages of human development, either by extending the range of accumulated fossils or by adding to our stockpile of cultural evidence.

But there remain yawning gaps, huge blocks of time for which we have no evidence at all of human existence. For the time between *erectus* and Neanderthal, for example, we know almost nothing: we face a wall of darkness that extends for hundreds of thousands of years.

This is the sort of gap that anthropologists concentrate upon, determined to track the fossil they know must lie hidden, somewhere, in the earth. The frustrating thing about fossil-hunting, of course, is that we can seldom choose our discoveries. If we track one particular fossil, we're likely to find another—and most often the one we find raises more questions than it answers.

Many ancient bones simply refuse to fit into the evolutionary scheme we have built for man. They present us with tantalizing hints of something new, something unknown, something that may force us to rewrite our books on human prehistory. Or they provide us with some bare new fact that is facinating in itself but devoid of clues that link it into our evolutionary story. Such riddles haunt the men and women who search for man's buried past.

These are the mysteries of prehistory—frustrating and fragmentary puzzles that guard their secrets so jealously that we can't begin to know them despite our growing wealth of shiny new tools and complex methods of identification.

One of these puzzles is the Heidelberg mandible (Plate 31), a giant jaw that has had scientists mumbling in their sleep since its discovery in 1907. Near Mauer, Germany, is located a huge sand-pit dug against the side of great Pleistocene deposits that had given up a vast collection of ancient animal fossils. It seemed a likely spot in which to search for the bones of prehistoric man. Dr. Otto Schoetensack was one of those drawn to the site—so drawn, in fact, that he visited the sand-pit almost daily for a

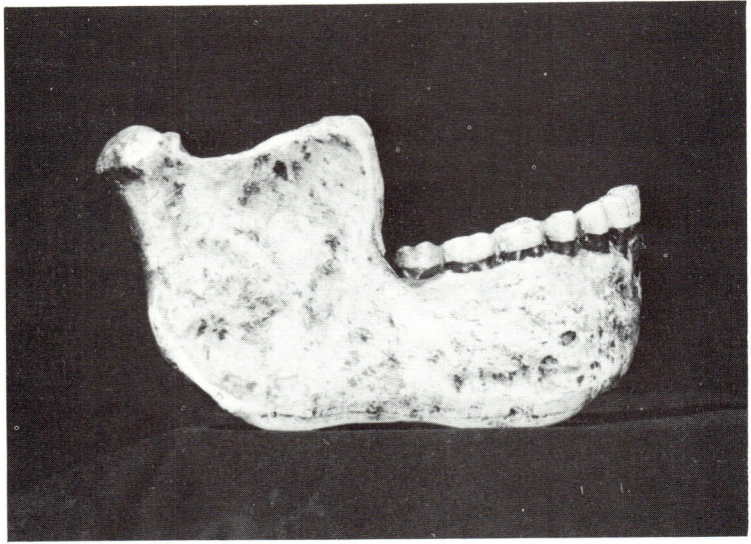

Plate 31. Cast of the Heidelberg mandible. *Photo courtesy Douglas H. Smith.*

period of more than twenty years. Finally, his patience was rewarded: a single immense human jaw emerged from a depth of some eighty-two feet.

The mandible was quite alone. Although the Mauer sand-pit yielded the bones of extinct elephant, bison, bear, and deer, none of these nestled in the sand adjacent to the jaw. Nor were there stone tools found nearby.

The jaw itself is huge, heavy, thick, and chinless. The teeth, still in their sockets, are small and worn flat like ours. They seem strangely out of proportion to the giant mandible in which they rest. Whoever owned this perplexing jaw in life must have been a curious mixture of primitive and advanced traits. But what is truly amazing is that the missing owner lived, geologists think, at least 400,000 years ago, a resident of Germany when hulking *Homo erectus* stalked the island of Java. How is it possible that a 400,000-year-old jaw features teeth almost as modern as ours? We don't know.

Three other fossils from Europe defy explanation. *Steinheim* (Plate 32) from Germany, represented by a single badly damaged skull lacking a lower jaw, was the victim of another plundering

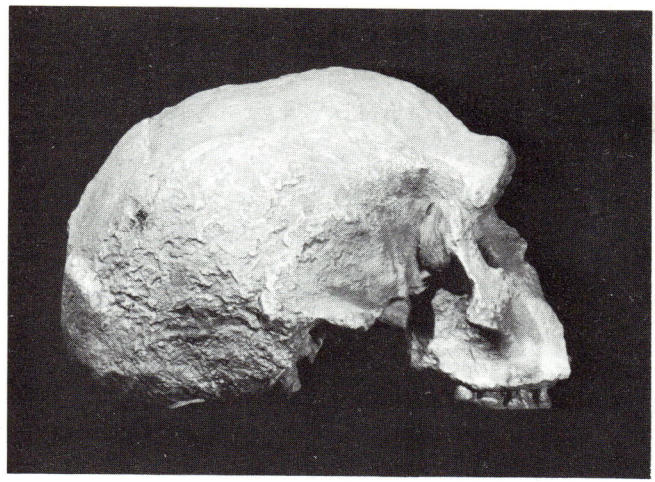

Plate 32. Cast of the cannibalized skull from Steinheim. *Photo courtesy Douglas H. Smith.*

cannibal. Found in a twenty-foot-deep gravel pit, the skull looks a little like Neanderthal, a little like *sapiens,* and a little like a mixture of the two. She may date from as early as 200,000 years before the present.

From Swanscombe Halt in England comes *Swanscombe* (Plate 33), represented only by the back portion of a female skull. The Swanscombe lady hoards her secrets, too. We know her age (250,-000 years) but little else about her.

Another riddle comes from France. Fontechevade man, known from two fragmentary skull fragments, is amazingly modern for his time (150,000 years ago) except that he is exceptionally thick-headed. His skull isn't so thick and heavy as that of Java man but surpasses in weight and density that known for any living variety of man.

These three fossils prove that a relatively modern braincase had evolved at a remarkably early time, closer in years to *Homo erectus* than to modern man. Can it be that "modern" types of men pre-dated Neanderthal, giving rise to both *sapiens sapiens* and *sapiens neanderthalensis?* Many experts think so. But they're stunned nevertheless to encounter so modern a skull form so near in time to Java man.

These are but three of the many fossils that continue to plague

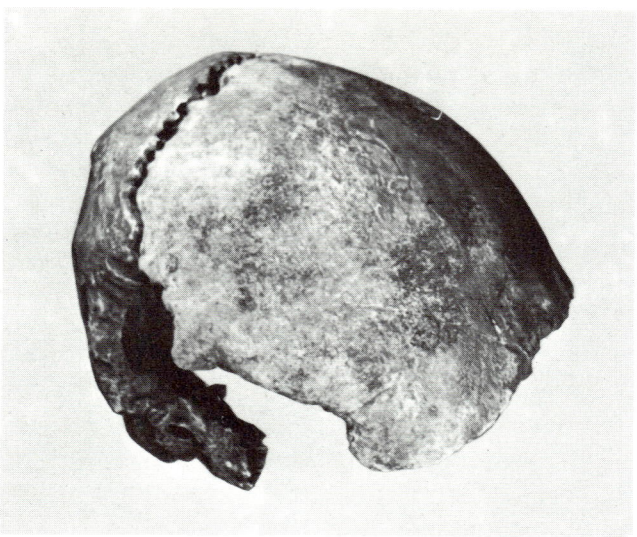

Plate 33. A female's skull from Swanscombe, England. Dated at 250,000 years, the skull is astonishingly modern in form. *Photo courtesy Robert Squier.*

anthropologists who'd like their fossils to fit nicely into a perfectly arranged history for man. And our list of such puzzles would extend to the moon and back if we were to include those about which the experts still haggle. Modern techniques of dating and anatomical analysis, however, have resolved our doubts about most of the other controversial finds. Most anthropologists have managed to agree for the most part on interpretations that satisfy them—at least until further evidence appears to allow fuller answers.

Individual fossils, however perplexing, are but a part of the immense mystery of prehistory. Even if it were possible to learn the identity of each extinct population, it would not be enough for us. We want to know how each form lived, where it came from, and where it went and why. We'd like to know how each fossil group dealt with the problems of everyday life, how early man fended off the bitter cold or coped with staggering heat, how he learned to fashion his weapons and battle his enemies, how and why he crossed deserts, mountains, and vast bodies of water. What did he think of his gods? How did he appease them? Did he cringe from death or, like the ancient Egyptians, look upon it as a joyous journey into a better world? What did he do for a head-cold, for a broken limb, for a cut finger? How was his social life, in those days before men dreamed of mouthwashes and deodorants and books of etiquette? Did he long for the status of chieftain? Did he have time for games and sports? Did he beat his wife? Did he ponder the origin of the men who passed before him?

All of this is lost in time. But clues to some of these mundane mysteries appear again and again—and in the strangest places.

Because poor health has no rival as man's greatest dread, some scientists concentrate their energies on discovering the means by which man treated his wounds and injuries. Studies of modern primates tell us that the earliest human populations treated their wounds instinctively, licking and sucking at the site of pain to relieve discomfort. But magic was an early invention. We know from ritual objects found in Pleistocene caves and from observations drawn from contemporary primitive peoples that sympathetic magic was put to quick use as a means of combating injury and illness. Did an arrowpoint penetrate a human body? Treat the wound—and treat the arrowpoint, robbing it of its fatal magic. Is a man plagued by jaundice? Treat it with juices extracted from the leaves of a yellow plant. We can imagine, in very ancient times, a growing specialization among certain gifted tribal members for the treatment of disease—sorcerers and shamans prescribing treatment through mystical incantation and exorcism.

We know that many primitive cures consist of simple common-sense methods. Our native Indians favored poultices, herb potions, massages, purgatives, sweat-baths, even enemas, for the alleviation of various discomforts. And every known primitive group boasts a huge stockpile of drugs extracted from seeds, roots, bark, and minerals. Many of these have found their way into modern pharmacology. We must expect that prehistoric man, accustomed to fending for himself, exhibited the same common sense. Was a man feverish? Apply leaves moistened with cold water. Was he chilled? Wrap him in warm hides and skins. Did he break a leg? Minimize pain by packing in clay, which hardened to form a protective cast; add crude splints for stability.

From various cave paintings, we know that by late Pleistocene times, man had learned something of anatomy: the animals he painted there are depicted often with hearts exposed and blood trailing. And for a man whose livelihood depended upon the capture and butchering of game animals, it seems likely that he would soon learn the fundamentals of physiology and compare his own body structures with those of the animals he slaughtered for food. And so he must have begun early to experiment with simple, crude surgery. The fine stone blades he so carefully fashioned must have been used to extract splinters and other foreign bodies.

But the most extraordinary surgery of prehistoric time is *trepanation,* the surgical removal of portions of the living skull. As early as 3000 B.C., man dared to open the skull for magical or medical purposes. Human skulls have been found all over the world—in Africa, Asia, Europe, North and South America, and the Pacific—which bear the marks of this remarkable skull surgery. From Peru especially come amazing examples of the daring of prehistoric surgeons (Plate 34). What is astonishing is that so many ancient patients survived the dangers involved in permitting a surgeon to cut out, with only a crude stone knife, a huge circle or square of living skull bone: of the trepanned skulls recovered from archaeological sites, more than half show healthy signs of new bone growth around the cut. And several skulls bear the scars of not just one healed trepanation but as many as *six* separate operations!

Why did ancient peoples risk such dangerous and radical surgery—and what accounts for their amazing success? We will never know for certain: all we have are these incredible skulls in which great artificial openings have been cut. Perhaps the operations were performed to alleviate the pain of headache or skull fracture. Or did they, like blood-letting (which must also have been a prehistoric treatment), permit the release of evil spirits trapped inside the skull—spirits which, left unexorcised, would cause convulsion or coma or vertigo?

Another mystery surrounds the population of the New World. We know that primitive hunters entered North America through the Bering Strait some 20,000 to 40,000 years ago and filtered down through Central and South America to inhabit all parts of the New World. But archaeologists working in Central and South America have encountered strange hints of prehistoric visitors

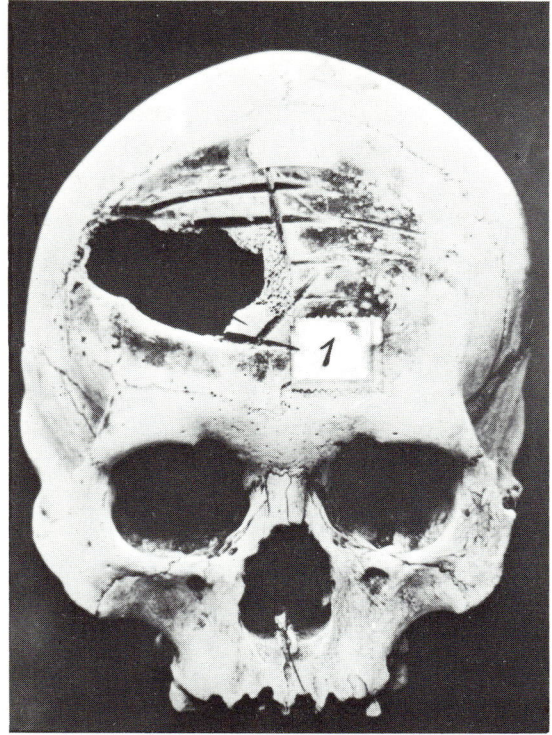

Plate 34. Trepanned skull from Peru. *Photo courtesy The Smithsonian Institution.*

from the Old World—men who journeyed across the oceans in the days before ships are known to have existed. Some anthropologists believe that they have found unmistakable evidence of early Egyptian contact: sun-worship cults, skilled techniques of human mummification, particular styles of hieroglyphic writing, pyramids and monuments—all of which, some think, were carried here by ship centuries before the time of Christ.

It is true that the New World is filled with archaeological wonders: vast ceremonial centers with great temples, sky-scraping pyramids, and ball-courts astronomically oriented and obviously planned by powerful central authorities. In Peru at the time of the Spanish conquest were aquaducts, paved roads, suspension bridges, and peoples whose calendars were more exact than our own. Some experts believe it impossible that the primitive bands of hunters who entered the New World through the Bering Strait could have erected, in so short a time, the splendid centers of civilization known to have existed in Central and South America or achieved such heights of cultural sophistication. They reject the hypothesis that such cultural duplications as hieroglyphics or methods of mummification known in both the Old and the New Worlds could result from independent invention. Instead, they believe that, at some unknown time in the past, Old World adventurers managed to span the oceans to bring knowledge and culture to the aborigines of the New World and to help build the extraordinary cities of the Aztecs, Mayas, and Incas.

Other experts argue that there could have occurred no such importations of peoples across the seas in pre-Columbian times. The oceans were then uncharted, the only known vessels clearly unseaworthy, the hypothetical voyage too long to sustain peoples with paltry food supplies.

But it is possible that occasional Old World contacts were made with the Indians of Central and South America—an accidental storm-driven crossing by a durable early craft or a chance-in-a-million lucky landing by precocious explorers bent on finding gold or slaves in a distant, unknown land. The question of trans-oceanic contacts rages as one of the most controversial among those who hope to recapture from time's grasp the story of man and his migrations. And the thought of such contacts, however unlikely, is one of the most exciting in prehistory.

What of Stonehenge, the group of vertical and horizontal stones located in England that represent some ancient and unknown temple of an equally unknown religion. Some of the stones come from a quarry in Wales, more than 150 miles away. How were they carried to Salisbury Plain by prehistoric men who'd never seen an air-lift or visualized a railroad? No one knows, but the first earthwork at Stonehenge dates from almost 2,000 years before Christ. The erection of this gigantic circle of stones must be counted as one of the most remarkable architectural achievements in prehistoric Europe.

So, too, must the amazing great stone spheres of Mexico. Archaeologists have uncovered here hand-worked balls of solid stone, some eleven feet in diameter and weighing more than twelve tons. Who made them? Who transported them and how?

What mysterious ritual impulse drove men to attempt such inexplicable sculpture 3,000 years before the discovery of the New World?

And what about the giant *moai*—the great stone faces of Easter Island? Twelve feet high, weighing some twenty-five tons, these immense statues litter the side-slopes of Rano Raraku volcano. Carved from tuff (a rock formed of cinders and volcanic ash), they date from 1,300 years into the past. Scientists estimate that each long-nosed, long-eared, legless statue must have taken at least a year to sculpt by teams of men working around the clock—and this on an island that could not have supported more than a few thousand inhabitants at any one time. Modern investigators, attempting to raise the toppled statues, were forced to use giant cranes to lift them. How did the missing sculptors accomplish this feat in prehistoric times? Why do the present-day natives know nothing of their ancestors? And what is the meaning of the other Easter Island mysteries—the curious elevated tombs, the carved but indecipherable written tablets, the hundreds of ancient stone towers?

Who were the inventors of the first mass-produced stone tools? Toward the end of the second interglacial period—perhaps 200,000 years ago—man devised a new technique of manufacturing flake tools by a special method of core preparation. By very careful trimming, he shaped a stone core to resemble a tortoise shell, then removed small flakes from the core's perimeter. Later, he reworked these rapidly produced flakes into a variety of scrapers and points. The invention spread quickly throughout Europe, the Near East, southwest Asia, and Africa. But scientists have never been able to find human bones in association with such tools.

And whose were the hands that fashioned the lovely, symmetrical laurel-leaf blades (Figure 9) and shouldered points called Solutrean tools? This delicately flaked tool technique is the most aesthetic—and the most mysterious—known from prehistoric times. Who were the stone-age hunters who labored so diligently over the manufacture of these extraordinary tools? Why did they devote such time and effort to the making of their implements? No human skeletal material has been found in any Solutrean site.

And speaking of missing fossils, why have we never found any fossil pygmy race? Pygmy populations abound in Africa, Asia, and

Figure 9. Laurel-leaf blade, an example of the remarkable parallel-flaking technique typical of the Solutrean culture.

Oceania—regions where larger fossils appear often to tell of previous human groups. No fossil skulls or skeletons have ever emerged to represent any pygmy population. Why is this fascinating group absent from the fossil record?

And what about Neanderthal? Did he lose out in a struggle for space and food with more modern types of man—or melt, through intermixture, into that new type? Or does he continue to exist today? Some Russian scientists, convinced that Neanderthal bands retreated and hid themselves in a desperate attempt to survive, regularly pursue the legendary "Abominable Snowman" in the hope that this creature does exist—and that he represents a hermit-descendant of the vanished Neanderthals. Stranger things have happened: the last North American Indian to endure in the wild did so by hiding in dense forests near densely populated California cities—until he was discovered in 1911, half-starved, foraging for food.

Why did no apelike or manlike forms evolve from the primitive proto-primates of the New World? Why is it that our distant nocturnal ancestors gave rise to descendants who are diurnal (active by day)? Why is man alone the "naked ape"—the only member of the superfamily *Hominoidea* to shed his hairy coat (long before the invention of clothing)? Why did the evolutionary development of early man's limbs precede that of his skull?

These are but a few of the mysteries that remain unsolved. Some are desperately important to our understanding of human history. Others simply nibble at our natural curiosity about the past.

All are questions about which scientists continue to ponder.

Most will never be answered.

Many may be but a single excavation away: Tomorrow a shovel may uncover the single thread that will unravel a host of Pleistocene riddles.

That is why we keep digging.

9

Journey into the Future

To track fossil man is to wield a pick and shovel, to lift rock and move rubble, to crawl head-first into dark and slimy caves, to dig and dig and dig some more despite bashed fingers and tired, aching muscles—all without the slightest guarantee that a single fossil will emerge from the ancient soil.

Any anthropologist will tell you, of course, that field-work has its own small rewards. While on a dig, the anthropologist is free to grow a huge, scraggly beard or chat noisily with himself without disgracing his family. But these simple pleasures are measured against the loneliness and inconvenience of life in the wild. To hunt for ancient bones is to live for much of the year at an isolated site where snakes and spiders abound. It is to celebrate holidays alone with a cold can of baked beans, a salt tablet, and a cup of gritty coffee. It is to itch and dream longingly of hot baths and clean sheets. It is to learn to live with frustration and, often, without hope of success. No wonder the anthropologist occasionally finds himself wondering why, after acquiring an advanced education, he devotes his life to manual labor under conditions no field-hand would tolerate.

Then one crumbling fossil comes to light beneath the spade.

And in an instant the itching and sweating and digging and waiting are all forgotten: the anthropologist is off to plead for funds that will finance another six months in the field.

Scientists track a fossil population for the same reason that mountain climbers scale a peak: because it is there. Fossils are not nearly so visible as mountain-tops. But they *are* there for the taking—if only we have the time and skill and patience and sheer good fortune to select and survey the right site. What other quest holds the drama and excitement of plucking from the distant past a bone never seen by human eyes—or gazing at artifacts left behind by a vanished race?

But there are other reasons for digging. However fascinating the bones of extinct animals and men, these are of little use to us unless we use them to answer modern questions and solve modern problems. Science earns its importance by applying the data it collects.

So what must we do with our stacks and heaps of accumulated fossil bones? We group them in temporal sequence from ancient to modern, learning how evolution operated in the past to produce today's living plant and animal populations. Such an arrangement of fossils, correlated with our knowledge of past environments, helps us to understand the intricacies of *ecology,* that delicate balance between each living organism and its surroundings. Once we know these, we can work efficiently to save endangered species, to promote conservation, to combat pollution, to halt the reckless squandering of precious natural resources—all to keep ourselves from going the way of those unfortunate dinosaurs.

Knowledge of past evolutionary trends and on-going evolutionary principles is put to use in breeding superior plants and animals, assuring food for our growing population. Primitive man first practiced agriculture and animal husbandry in a crude, unconscious attempt to improve upon nature. Today, we utilize complex procedures based upon scientific knowledge of genetics. But the aim is the same: to breed for the best qualities in both plants and animals. We produce today the largest and finest of food plants, the leanest and most nutritious of food animals. Remember, too, that the rose by your window and the pedigreed pup at your side are products of man's ability to apply the principles of evolution for his own comfort and pleasure.

Anthropologists who concentrate on prehistoric human populations come to know how past groups adapted to their individual environments. Anthropologists want to know how people survive under rigorous conditions so that we can take full advantage of all habitable world areas; this is another measure demanded by our growing population. Studies of both living and fossil groups, for example, help us to understand how the inhabitants of the Andes have adapted to high-altitude living; how the natives of Tierra del Fuego survive its awesome cold; how the peoples of the world's deserts have adjusted to almost intolerable heat. We want to know too—especially from fossil hunters and gatherers—how primitive man managed to exploit seemingly barren landscapes, what sorts of food he wrested from meager vegetation, how he utilized animal products to help sustain life. All these facts help us devise means to feed, clothe, and supply space for an expanding world population. And such facts tell much about human nature. These insights are necessary if we are to bring about social change and shape a strong and hopeful future.

One of the most urgent concerns of anthropology is the matter of human race. Ours is a world torn by racial conflict and misunderstanding—and there can be no better time for the anthropologist to contribute his skills and knowledge in helping to alleviate that conflict. Who better understands how races evolve, how they differ, and how those differences affect the lives of men and women of different colors?

Hoping to put into proper perspective the problems of race, social scientists joined with physical anthropologists to provide the information on which the UNESCO Statement on Race was based. This statement makes clear that all men are derived from the same common stock, that all men belong to a single species, and that differences that exist between different human groups result from the operation of complex evolutionary factors. It emphasizes too that so far as temperament, personality, character, and intelligence are concerned, there exist no inborn differences between races. No single race has a corner on brain-power, or moral fiber, or artistic talent.

The anthropologist does *not* say that everyone is born equal. He knows as well as the rest of us that there are individuals of every race who are superior in one quality or another to other individuals. What the anthropologist does say is that human *races* are equal: that no race as a whole is smarter or more creative or talented than any other. Such differences are personal, varying from one individual to another and not from race to race. If some human groups are culturally more advanced than others, this advantage arises from greater opportunities and not from some innate superiority. These are the findings of those who seek out fossil man.

Only one who knows the evidence can explode racial myths.

Only one who devotes much of his life to understanding race can adequately argue—with proof—that racial equality is a fact and not a goal. It remains for the anthropologists to pass along these judgments to social scientists so that the latter can implement these findings and this proof to the betterment of all mankind.

If race is an agonizing problem in most nations, disease is a plague that stalks man in every land and at every time. The study of ancient disease permits us to combat the effects of illness in the present and in the future. Some anthropologists specialize in paleopathology, a field so wide that it boasts a dozen or more sub-disciplines. One of these is disease migration. People are more mobile today than ever before. The emergence of new nations, the placement of armed forces and advisory personnel in remote regions, the venture of private enterprise into unfamiliar lands, the new affluence that allows individuals to travel all across the globe—all these are factors that increase contact between peoples previously isolated. With increased mobility comes the spread of disease. If we can learn how disease is transmitted and how past peoples dealt with new and previously unencountered diseases—we can know how we must now attack new and recurring afflictions.

Tracing ancient patterns of disease makes it possible for us to find the causes of some modern diseases. Sometimes, the findings of anthropologists alter the direction of medical research. Arteriosclerosis (hardening of the arteries), for example, was believed to occur as a result of the modern use of tobacco and alcohol—until this condition was found in a 1225 B.C. mummy of an Egyptian pharaoh; the mummy came from a time and place where tobacco was unknown and alcohol rarely used. Fossil studies show also that bacterial infections were rare until man began to group together in cities. We can't abandon our cities, but we can direct medical research toward the alleviation of contagion.

Dental studies too benefit from skeletal collections. At about the time that man first domesticated plants (thus trading his protein diet for a carbohydrate one) he was increasingly susceptible to dental decay. Once this was known, dental researchers could concentrate their efforts on learning how diet affects dental health. Anthropologists noted also that where native waters are high in natural fluoride, ancient skeletons display an amazingly low incidence of dental caries—a fact that has our governmental officials arguing for the addition of fluorides to municipal water supplies.

Knowledge and skills derived from our studies of fossil life provide the basis for sciences of the future—among them, *exobiology,* the study of extraterrestrial life. For there are frontiers left to conquer: When man first set foot upon the moon in July, 1969, he challenged the universe. Unexplored worlds await him— each to be probed for its unique history.

And the earth? It still holds secrets and puzzles untapped. Man is driven by the need to explore and to understand. The same impulse that has sent him around the world to grub energetically in the earth's surface and into outer space to probe other planets has led him in the last century to "inner space"—to the sea, where the new field of underwater archaeology promises exciting new discoveries.

The first, groping attempt to find ancient man underwater came in 1885, when an American named E. H. Thompson determined to explore the Sacred Well of the Mayas on the Yucatán peninsula. Thompson had read of the sacred *cenote* at Chichén Itzá, where, long before the discovery of the New World, ancient Mayan priests threw sacrifices into this black and bottomless pool—not only objects of jade and beaten gold but also human sacrifices, beautiful young maidens and heroic Mayan warriors. Other archaeologists had pondered the legend of the *cenote*. But it was, they argued, only a legend. Spanish *conquistadors* had looted the treasures of Maya-land. What could be left? Besides, the well was inaccessible, littered with boulders, rotten trees, leaves, and stagnant fungus.

But Thompson longed to retrieve the treasure he knew must litter the floor of the well.

Unable to obtain financial backing for his risky expedition, Thompson proceeded alone. He traveled to Yucatán and viewed the immense pool that legend cited as the sacred *cenote*. Draining was out of the question, as was diving: debris centuries old lay stagnant in the well, and boa constrictors abounded in the encroaching jungle. Thompson erected a crude derrick, fitting a steel bucket to a swinging boom. And he set to work with native helpers, scooping endless buckets of muck and rubble from the well.

The days stretched into weeks. Thompson became discouraged. But then the bucket emerged from the depths of the well to disgorge its content one last time—and it yielded copal, fragrant Mayan incense. Soon more balls of copal emerged, then arrowpoints, broken pottery, obsidian figurines—and then golden bells mixed with the bones of human skeletons. Thompson had proved the legend of Mayan sacrifice to be true.

Afraid that his crude scooping method would miss objects even more valuable, Thompson dispatched divers, although the water was so dark and scummy that their lights were useless. Groping in the black depths, the workers reaped a rich harvest of Mayan treasures: delicate copper bells, rings of pure gold, exquisite jade beads—and more human skeletons.

Across the world archaeologists got the message: they hired divers or learned to swim in the deep themselves. And they began at last to explore the sea. The invention of scuba (short for "self-contained underwater breathing apparatus") gear enabled divers to penetrate to depths previously unexplored. Now, equipped with more sophisticated equipment, they search for sunken ships, buried cities, hidden shorelines—and still more fossil bones. An amazing wealth of information rests hidden beneath the waves; unimagined treasures wait to be reclaimed. And the work has just begun.

There are other ways to know prehistoric man. Modern anthropologists pursue what they call "living prehistory"—a term that covers dozens of exciting new experiments and observations de-

signed to illuminate the distant past. Research on living groups—
often chimpanzees, baboons, gorillas, and orang-utans—supports
deductions drawn from evidence excavated from the earth. Such
men as Irven DeVore of Harvard University and John Crook of
the University of Bristol study intensively the lives of baboons in
the wild. Vernon and Frances Reynolds, Jane van Lawick-Goodall,
and others investigate the wild chimpanzees of Africa. These
adventurous scientists devote themselves so fully to their task that
many of them eventually are "adopted" into the primate troop
they seek to study. After a long period of observation—which gives
the primates time to become accustomed to the presence of interested humans—the scientists are permitted to join with the troop,
to follow as the animals roam through the forests, to eat with
them, to romp and play, and to learn to interpret the various
primate gestures.

Early man, of course, was neither baboon nor chimpanzee. But
studies have shown that free-ranging primates, like man, live in
organized groups with recognizable social structures. A good understanding of primate life gives us insights into the probable
social life of early man.

For a long time, anthropologists concentrated upon observing
artificial primate colonies—those that live in our zoos. They've
recently realized, however, that because zoo life differs radically
from life in the wild, the behavior of a zoo animal may be abnormal compared to free-ranging life for that same animal. So
science has taken to the jungles for information—and returned
with important new evidence regarding the life of early man and
of modern primates. Mrs. van Lawick-Goodall, for example, was
present to film the first documented incidence of chimpanzee
tool-making in the wild.

Field-work permits the observation of animals as they live undisturbed by man. Laboratory observation, on the other hand,
allows scientists to conduct extensive psychological studies with
various primates. Many tests with laboratory monkeys have led to
studies in normal and abnormal behavior for man. Experimenta-
tion at various primate centers, for example, has shown that young
monkeys separated from their mothers and reared without maternal affection often behave abnormally when they are grown;
monkeys given a "surrogate mother"—a soft doll or cuddly

blanket—behave more normally than those reared in the absence of either real or surrogate mothers. Such results have much to say about the importance of child-raising among humans.

Experiments with human infants yield additional information regarding instinct, sense perception, learning patterns, and communication and give rise to vast new testing possibilities—all with a view to solving problems of modern life.

Other anthropologists seek to know more about prehistoric man by testing themselves. Dr. Leakey has taught himself to make stone tools; by using these in butchering wild game, he learns how early varieties of men went about this chore. Sometimes he takes the role of a prehistoric hunter. Stalking game, he has learned—just as early man must have learned—to observe the movement of life around him, to detect the almost invisible presence of food animals, to pick up the trail of roaming game. Like many others, Dr. Leakey is learning by doing. Such men represent a new breed of anthropologists.

Every year "dead" and "living" evidence reveals more of man's buried past. But what of his future? How will man look in A.D. 1,000,000?

Some argue that man controls his own destiny, that civilization and advanced technology will cancel out the effects of evolution. But this is an absurd suggestion. Evolution operates just as surely on man today as it has in the past and as it does on all other members of the animal kingdom. Man has just begun to evolve: compared with other animal species, he is a young and untried animal.

Most agree that evolution will continue to shape man's future, although few risk guessing the form of man to come. A few science-fiction minded experts predict that man will become better formed and more functional, with bones grown so strong that they are unbreakable, superhuman muscles invulnerable to stress or strain, and eyes capable of X-ray vision. Future man, they think, will be freed from disease, aging, and death; future man will be Superman.

Others insist that man's future lies in the continuing development of his remarkable brain. Imagine a future race of man characterized by huge, bald heads, tiny little legs, and immense index fingers (for pushing all the buttons that will do his work for him, of course!).

Neither Superman nor Superbrain sounds particularly appealing. No need to worry: neither is likely to appear. Man has come this far because he is first of all a *generalized* animal, able to adapt with relative ease to diverse situations and environments. We don't think man will suddenly evolve such complex and astonishing specializations as X-ray eyes or shatter-proof bones.

Nor will his head get much larger because it's already large enough—relative to body size—to complicate child-birth. It is true, however, that man seems to be growing ever more round-headed, a trend apparent among numerous human populations at present.

Man may get to be a bit taller on the average as a result of improved diet and medical care—but he'll never tower over a giraffe. Man is a rather large animal to begin with, and his backbone—originally designed to serve a pronograde animal—is too fragile to haul about a huge, hulking body. Man's spinal column, in fact, is a rather makeshift arrangement, prone to slipped discs and wrenched vertebrae. If anything is to be done with it, we predict it will grow thicker and stronger—but not longer.

Judging from past evolutionary trends, it's relatively safe to make a few other predictions. The wisdom teeth will have to go—there's no room for them in the jaw as it is. Most of us suffer through the agony of having them pulled or cut out. Other lucky individuals are born without them. And this congenital absence of the third molar—appearing with increasing frequency—foretells the end of the wisdom tooth. And good riddance.

We'll probably make do in the future without that little fifth toe, too. It's a degenerate organ, short and squat and quite worthless as human toes go. We'd never miss it—although it will be a job for future anthropologists to rewrite all the taxonomic textbooks that classify man as a five-toed primate. Future anthropologists may consider it a good bargain in the end, however; if they can distinguish between ape and human bones on the basis of toe number, they won't have to sit around pondering tooth-and-jaw form, as we must today.

Anthropologists used to stress the fact that man is *not* the end product of evolution. When we see a neat arrangement of fossil forms, each transitional stage leading inexorably toward modern man, we sometimes get the notion that the whole game was rigged to produce man, king of all the beasts. But evolution doesn't work that way. No directions have been preordained, no unique forms deliberately shaped. Evolution has proceeded so far in random fits and spurts, discarding unsuitable forms as often as it created new ones.

But those same anthropologists are beginning to fear that modern man may prove after all to represent the last stage in human evolution—not because the evolutionary process followed predetermined paths, but because man is the first form to tamper with his own destiny.

Man, having gained the power to manipulate his own environment, possesses too the power to pollute it.

Man, having come to dominate nature, has failed to live in harmony with it.

Man, having devised the means to limit his own death rate, has not seen fit to alter his own birth rate.

Early man lived, like his nonhuman relatives, in balance with his environment. Life in the primitive world was both short and risky. Man competed with fierce animal predators for food and space. He fell frequent victim to disease and injury. Drought and famine helped to limit the size of human populations. Starvation was commonplace.

With the development of culture, however, man began to exercise some control over his own destiny. About 8,000 years ago, he unlocked the secret of agriculture, which permitted him to settle in permanent villages and abandon the riskier nomadic life. Although he faced new dangers associated with urbanization—among them, organized warfare and disease contagion—his existence, on the whole, was more secure. He began to stockpile food supplies against future famine. He diverted rivers in order to assure himself of water through dry periods. He learned the merits of order and organization, utilizing these in cooperation with his fellows against outside threats. He traded magic for science, learning the rudiments of medical practice.

In time, he built huge cities devoted to the perpetuation of the human race. Industrialization gave him added security. Population growth boomed. With the development of improved medical and surgical techniques, the birth rate soared—and the death rate slowed. There were more people on earth, both young and old, than ever before. Dazzling refinements in agriculture, including the introduction of powerful pesticides, permitted an ever-increasing population—and industry, hungry for labor, demanded it. By 1900, Man the Culture Bearer, most successful of all biological species, reached his prime. There seemed nowhere to go but up.

And up he continued to go. But in his rush to make things ever bigger, ever faster, and ever better, man forgot to keep track of the by-products of his own success. He dumped his industrial wastes into rivers, lakes, and oceans, spreading pollution there like an awesome cancer. He exploited the land, sacrificing millions of acres of precious earth to erosion. He contaminated every living animal population of the world with DDT. He multiplied himself in staggering proportions, unaware that his rate of reproduction far outstripped his ability, even with an advanced technology, to feed an exploding population. The world seemed huge enough to support an unlimited amount of people and pollution.

Now no animal population on earth is free of man's disruptive influence. Man faces today the most critical challenge of all: the salvation of his own environment. Will he watch helplessly as the quality of life deteriorates until, in the end, life itself cannot be supported on earth? Or will he accept the role thrust upon him by his own evolutionary success—that of caretaker to a planet he has plundered in the past?

If man is to survive for future evolution to act upon, he must face now the hard realities of pollution on land, in the water, and in the air. He must know the consequences of his use of chemical additives. He must recognize the futility of overpopulation. He must learn to respect the worth of animal forms less advanced than himself. He must relinquish the comforting but false hope that technological progress, quite inevitably, can solve the problems man creates through his careless exploitation of the world about him. Above all else, he must sacrifice rhetoric for action, tokenism for concerted effort—and he must do it *right now.*

He begins at home to repair the damage by refusing to use detergents that pollute the waterways, by exchanging his shiny new pollutant-belching automobile for smog-free air, by limiting the size of his own family. And then he must demand of others that they follow suit. It is not an easy role. Man, who so blithely accepts his place as king of beasts, is finding that the crown rests uneasy upon his brow.

Will man survive?

Some experts are pessimistic. They are angered by man's capacity for thorough and thoughtless destruction. They despair over his disregard for the welfare of future generations. They wring their hands over man's enthusiasm for "progress" at any cost.

But others are hopeful. They know that we have today the technological means of reducing human birth rates and increasing food supplies, and that we are rapidly developing means of reducing pollution. More important, there looms on the horizon a new generation, impatient with the excesses of the past and anxious to make environmental amends. Our journey into the future will be guided by their hands.

Glossary

ADAPTATION: the process by which an organism adjusts to its surroundings or way of life; the fitness of an animal or plant species to its environment.

ANTHROPOID: manlike; distinguishes the apes (gorilla, chimpanzee, orangutan, gibbon) from the monkeys.

APE: common name of the family Pongidae.

ARBOREAL: of or like a tree; adapted for living in a tree.

AUSTRALOPITHECINE: an early Pleistocene hominid taxon including the fossil man-apes (*Australopithecus*).

BRACHIATION: movement through the trees by swinging with the arms from branch to branch.

BRECCIA: a mass of material (earth, fossils, sand, fragmentary stone) that has become consolidated by a cementing matrix such as lime salts.

BURIN: a flake tool for sculpting or engraving, associated especially with Old World Upper Paleolithic cultures.

CARBON-14: a radioactive isotope of carbon, useful for estimating the age of objects preserved more than 70,000 years.

Chellean: term derived from site at Chelles, France, which designates that division of the Paleolithic during which certain flint tools were made; patterned flint tools.

Dendrochronology: a method of dating by means of tree rings.

Diastema: a space between the teeth; refers to a gap in mandible or maxilla to accommodate the protruding canine of the opposite jaw.

Dryopithecus: Miocene and Pliocene taxon of apes, represented by several species, ancestral to the living apes and perhaps ancestral to man.

Ecology: the balance of relations of organisms with one another and with the surrounding environment.

Eutherian: an infraclass of mammals that nourish their unborn young by means of a placenta.

Evolution: the fact of gradual change in species of plants and animals through succeeding generations; descent with modification.

Fluorine Analysis: a technique for dating by means of amounts of the element fluorine absorbed in fossil bone.

Foramen magnum: the large opening at the base of the vertebrate skull through which the spinal cord enters the brain.

Gene pool: a breeding population.

Gigantopithecus: a large fossil ape from China.

Glaciation: the formation of ice sheets on land.

Graver: a flint tool used as a chisel.

Hominid: pertaining to the family Hominidae, or man, and thus referring to characteristics deemed "manlike"; not to be confused with *hominoid,* which refers both to the apes and to man.

Insectivore: an insect-eating mammal; a group closely related to the early Mesozoic mammals.

Ischial callosities: patches of tough, fibrous tissue overlying the ischial bones of the pelvis, seen in some apes and most Old World monkeys.

Mandible: lower jaw.

Maxilla: upper jaw.

Metatherian: an infraclass of mammals containing the marsupials, or pouched animals.

Mousterian: a varied complex of stone-tool industries so far exclusively associated with Neanderthal man.

Mutation: a permanent change in a gene carried by an individual; this change can be transmitted to future generations.

Occipital: pertaining to the back of the head; the rearmost bone of the skull, penetrated by the *foramen magnum.*

OLFACTORY: pertaining to the sense of smell.

OPPOSABLE: refers to the first digit (thumb or big toe) of typical primates, meaning an ability to rotate this digit and permit it to be opposed to other digits in grasping functions.

PARANTHROPUS: now called *Australopithecus robustus,* refers to the species *Australopithecus,* which is now extinct.

PARIETAL: a bone that forms the "wall" of the skull vault; the parietals occur in pairs.

POLYCHROME: many-colored.

POLYTYPIC: occurring in several distinguishable breeds; *Homo sapiens* is a *polytypic* species because it contains many types of man.

PONGID: member of the family Pongidae, as opposed to hominid; also refers to "apelike" characteristics, those anatomical traits (e.g., heavy brow, projected canine) reminiscent of the great apes.

PREHENSILE: adapted for seizing or grasping, as the prehensile tail of many New World monkeys.

PROGNATHISM: forward protrusion of the jaws.

PROTOTHERIAN: an infraclass of mammals containing the monotremes (for example, the duck-billed platypus).

RACE: a population within a species, distinguished from other such populations on genetic bases.

RADIATION, ADAPTIVE: the evolutionary branching out from a basic animal form.

SIMIAN SHELF: a bar of bone between the two sides of a mandible; seen in apes and Old World monkeys.

SPECIALIZED: an animal or organ adapted to a particular environment or way of life; the human foot, adapted for upright stance, is a specialized organ.

SPECIES: the basic taxon among bisexual organisms; a population that interbreeds to produce fertile offspring.

STEREOSCOPIC VISION: depth perception made possible by fusion, in the brain, of images seen by two eyes located so closely together that their fields of vision are almost the same.

TAXON: term of classification designating a group of animals related by common descent.

TAXONOMY: the science of systematic classification of plants and animals in a manner that reflects their relationships with one another.

TREPANATION: the surgical removal of cranial bone; also termed trephination.

VILLAFRANCHIAN: an assemblage of animals whose appearance marked the beginning of the Pleistocene; named after a site in Italy.

VIVIPAROUS: bearing living young instead of laying eggs.

For Further Reading

Chapter 1. THE MONKEY WAR

Darwin, Charles *On the Origin of Species* (London: John Murray, 1859)
Eiseley, Loren *Darwin's Century* (London: Gollancz, 1959)
McKern, Thomas W. and Sharon S. McKern *Human Origins* (Englewood Cliffs, N.J.: Prentice-Hall, 1969)

Chapter 2. THE ANCIENT WORLD

Brothwell, D. R. *Digging up Bones* (London: British Museum of Natural History, 1965)
Moody, Paul Amos *Introduction to Evolution* 2nd ed. (New York: Harper & Row, 1962)
Oakley, Kenneth *Frameworks for dating Fossil Man* (London: Weidenfeld & Nicolson, 1964)
Simpson, George Gaylord *Life of the Past* (New York: Bantam Books, 1968)

Chapter 3. THE PRIMATE PATH TO MAN

Clark, W. E. LeGros *The Antecedents of Man* (New York: Harper & Row, by arrangement with Edinburgh University Press, 1959)

Morris, Ramona and Desmond Morris *Men and Apes* (New York: Bantam Books, 1968)

Schaller, George B. *The Year of the Gorilla* (London: Collins, 1964)

Chapter 4. MAN-APES AND APE-MEN

Clark, W. E. LeGros *Man-Apes or Men-Apes?* (New York: Holt, Rinehart & Winston, 1967)

Leakey, L. S. B. and Vanne Morris Goodall *Unveiling Man's Origins* (New York: Putnam, 1969)

Morris, Desmond *The Naked Ape* (London: Cape, 1967)

Chapter 5. MAN EMERGES

Braidwood, Robert J. *Prehistoric Men*, 7th ed. (Chicago: Scott, Foresman, 1967)

Howells, William *Mankind in the Making* (London: Penguin, 1967)

Koenisgald, G. H. R. Von *The Evolution of Man* (Ann Arbor: University of Michigan Press, 1962)

Chapter 6. THE NEANDERTHALS

Boule, Marcellin, and Henri V. Vallois *Fossil Men* (London: Macmillan, 1957)

Quennell, Marjorie, and C. H. B. Quennell *Everyday Life in Prehistoric Times* (London: Batsford, 1959)

Chapter 7. MONSTERS, RACES, AND MODERN MAN

Bushnell, G. H. S. *The First Americans* (London: Thames and Hudson, 1968)

Chard, Chester S. *Man in Prehistory* (New York: McGraw-Hill, 1969)

Jennings, Jesse D. and Edward Norbeck (eds.) *Prehistoric Man in the New World* (Chicago: University of Chicago Press, 1964)

Chapter 8. MYSTERIES IN PREHISTORY

Bushnell, G. H. S. *Peru* (London: Thames and Hudson, 1966)

Bydoux, Henri-Paul *The Buried Past* (London: Weidenfeld & Nicolson, 1966)

Hawkes, Jacquetta (ed.) *The World of the Past* (London: Thames and Hudson, 1964)

Chapter 9. JOURNEY INTO THE FUTURE

Mead, Margaret, *et al.* (eds.) *Science and the Concept of Race* (New York: Columbia University Press, 1968)

National Geographic Society *Vanishing Peoples of the Earth* (Washington D.C.: National Geographic Society, 1968)

Stern, Curt *Man's Genetic Future* in Thomas W. McKern (ed.) *Readings in Physical Anthropology* (Englewood Cliffs, N.J.: Prentice-Hall, Inc., 1966)

Index